Jo Marchant is an award-winning science journalist. She has a PhD in genetics and medical microbiology from St Bartholomew's Hospital Medical College, London, and an MSc in Science Communication from Imperial College, London. She has worked as an editor at *New Scientist* and *Nature*, and her articles have appeared in publications including the *Guardian*, *Wired UK* and the *Observer*, the *Economist* and *Smithsonian* magazine. She has lectured around the world. In 2009, *Decoding the Heavens* was shortlisted for the Royal Society Prize for Science Books, and in 2016, *Cure* was shortlisted for the Royal Society Insight Investment Science Book Prize.

Also by Jo Marchant

Decoding the Heavens
The Shadow King

CURE

A JOURNEY

INTO THE SCIENCE OF

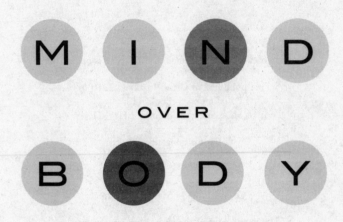

MIND

OVER

BODY

JO MARCHANT

CANONGATE

This paperback edition published in 2017 by Canongate Books

First published in Great Britain in 2016 by
Canongate Books Ltd,
14 High Street, Edinburgh EH1 1TE

www.canongate.co.uk

1

British Library Cataloguing-in-Publication Data
A catalogue record for this book is available on
request from the British Library

ISBN 978 0 85786 885 5

Typeset in Adobe Garamond and Avenir by
Palimpsest Book Production Ltd, Falkirk, Stirlingshire

Printed and bound in Great Britain by Clays Ltd, St Ives plc.

To my parents, Jim and Diana Marchant.
Thank you for teaching me to think,
question and explore.

CONTENTS

AUTHOR'S NOTE

Many scientists and patients shared their knowledge and experiences with me for this book. They aren't all directly mentioned in these pages, but I'm overwhelmingly grateful to each of them.

Quotes that are not referenced in the notes are taken from my own interviews with patients and practitioners. All referenced quotes are from interviews with me, or from other published sources, and these are flagged in the text with citations in the notes.

I have changed some individuals' names to protect their privacy – in these cases I refer to the person by a first name only. If a full name is given, that is the person's true identity. (Exceptions are Davide in Chapter 1 and Fhena in Chapter 10 – these are their actual first names.)

INTRODUCTION

One weekday morning last summer, I was in the local park. It was a cheerful south London scene, with kids splashing in water fountains and playing football on the grass. I perched on the edge of the sandpit with two other mothers, clutching sun cream and rice cakes as we watched our children build lopsided castles with brightly coloured plastic spades.

One of the women, a bright, articulate mum I had just met, was explaining how a homeopathic medicine had cured her of longstanding, debilitating eczema. 'I love homeopathy!' she said. As a scientist, I had to protest. Homeopathy is effectively water (or sugar pills) in fancy bottles – any active substance in these treatments is diluted far beyond the point at which any single molecule of the original could possibly remain. 'But there's nothing in homeopathic remedies,' I said.

My new friend looked at me scornfully. 'Nothing *measurable*,' she replied, as if I were slightly dim for not grasping that its healing properties are due to an indefinable essence that's beyond scientists' reach. And in those two words, I felt that she summed up one of the major philosophical battles in medicine today.

Stacked up on one side are the proponents of conventional, western medicine. They are rational, reductionist and rooted in the material world. According to their paradigm, the body is like a machine. For the most part, thoughts, beliefs and emotions don't feature in treatment for a medical condition. When a machine is broken, you don't engage it in conversation. Doctors use physical methods – scans, tests, drugs, surgery – to diagnose the problem and fix the broken part.

On the other side is, well, everyone else: followers of ancient, alternative and eastern medicine. These holistic traditions prioritise the immaterial over the material; people over conditions; subjective experience and beliefs over objective trial results. Rather than prescribing physical drugs, therapists using acupuncture, spiritual healing and reiki claim to harness intangible energy fields. Advocates of homeopathy aren't concerned that their remedies contain no physical trace of the active ingredient, because they believe that an undetectable 'memory' of the drug somehow remains.

Conventional medicine still has the upper hand in the west, but alternative medicine is embraced by millions of people. In the US, the wonders of spiritual healing and reiki are regularly discussed on television news. As many as 38% of adults use some form of complementary or alternative medicine (62% if you include prayer). Each year they spend around $34 billion on it,[1] with 354 million visits to alternative medicine practitioners (compared with around 560 million visits to primary care physicians).[2] In London, where I live, mothers commonly put amber necklaces on their babies in the belief that this gemstone has the power to ward off teething pain. Intelligent, educated women reject crucial vaccines for their children and, like my friend, embrace treatments that make no scientific sense.

Not surprisingly, scientists are fighting back. Professional sceptics on both sides of the Atlantic – debunkers like James Randi and Michael Shermer; scientist bloggers like Steven Salzberg and David Gorski; the biologist and author Richard Dawkins – aggressively denounce religion, pseudoscience and especially alternative medicine. The 2009 book *Bad Science*, in which epidemiologist Ben Goldacre slates those who misuse science to make unjustified health claims, has sold more than half a million copies in 22 countries. Even comedians from Tim Minchin to Dara Ó Briain are joining the fight, using their jokes to champion rational thinking and point out the absurdity of treatments like homeopathy.

Their followers are standing up against the tide of irrationality with meetings, articles, protests, and what science journalist Steve Silberman calls 'anti-woo lines drawn in the sand',[3] such as a petition signed by hundreds of UK doctors demanding that the National Health Service stops spending money on homeopathic treatments. Clinical trials prove that most alternative remedies work no better than placebos (fake treatments), the sceptics point out – people who use them are being duped. Many argue that these bogus treatments need to be stamped out. There's nothing we need in healthcare that we can't get from conventional, evidence-based cures.

I'm all for defending a rational world view. I believe passionately in the scientific method: I have a PhD in genetics and medical microbiology, during which I spent three years probing the inner workings of cells at a top London hospital. I believe that everything in nature can be studied scientifically if we ask the right questions, and that the medical treatments we put our trust in should be tested in rigorous trials. The sceptics are right: if we abandon science for wishful thinking we might as well be back in the dark ages: drowning witches, bloodletting and praying that God will save us from the plague.

But I'm not sure that simply dismissing alternative medicine is the answer. In my work as a science journalist, I encounter not just those who are cured by modern medicine but those who aren't: patients whose lives are devastated by gut problems or fatigue yet are dismissed as not having a 'real' condition; people suffering from chronic pain or depression, prescribed ever-higher doses of drugs that create addiction and side effects but don't solve the underlying problem; cancer patients who receive rounds of aggressive treatment well past the point at which there's any reasonable hope of extending their lives.

And I regularly come across scientific findings – sometimes making headlines but often buried in specialist journals – suggesting that intangible, immaterial treatments can have real physical benefits. Patients hypnotised before surgery suffer fewer complications and recover faster. Meditation triggers molecular changes deep inside our cells. And as we'll see in the first chapter of this book, if a treatment works no better than placebo that doesn't mean it doesn't work – simply believing you have received an effective remedy can have a dramatic biological effect. The mothers around me using amber bracelets and homeopathic pills aren't ignorant, or stupid; they know from experience that these things genuinely help.

So although I believe that the alternative medicine advocates are deluded with their talk of water memory and healing energy fields, I don't think the sceptics have got it completely right either. I started to write this book because I wondered whether they, along with conventional doctors, are missing a vital ingredient in physical health; an omission that's contributing to the rise of chronic disease and sending millions of sane, intelligent people to alternative practitioners. I'm talking, of course, about the mind.

***

Have you ever felt a surge of adrenaline after being narrowly missed by a car? Felt turned on just from hearing your lover's voice? Retched at the sight of maggots in the trash? If so then you've experienced how dramatically the workings of your mind can affect your physical body. Information from our mental state constantly helps our bodies to adapt to our surroundings, even though we might not be aware of it. If we see a hungry predator – or an approaching truck – our body prepares itself to get out of the way, fast. If someone tells us food is coming, we get ready for a nice, relaxing spell of digestion.

This much we know. Yet when it comes to health, conventional science and medicine tend to ignore or downplay the effect of the mind on the body. It's accepted that negative mental states such as stress or anxiety can damage health long-term (though even this was highly contested until a few decades ago). But the idea that the *opposite* might happen, that our emotional state might be important in warding off disease, or that our minds might have 'healing powers', is seen as flaky in the extreme.

The split between mind and body in western medicine is commonly blamed on French philosopher René Descartes. Ancient medics, with little to work with beyond the placebo effect, knew full well that mind and body were entwined. The early Greek physician Hippocrates, often described as the father of medicine, apparently spoke of 'the natural healing force within', while the second-century doctor Galen held that 'confidence and hope do more good than physic'.[4]

But in the seventeenth century, Descartes distinguished between two fundamental types of matter: physical objects, such as the body, which could be studied by the scientific method, and the immaterial, mental spirit, which he believed was a gift from God and could not be studied scientifically. Although these two forms of matter could communicate (Descartes thought this happened via the brain's pineal gland), he concluded that they exist independently. When we die and no longer have a body, our self-contained spirit lives on.

Most philosophers and neuroscientists now reject these ideas about mind–body dualism. Instead, they believe every brain state – each physical configuration of neurons – is intrinsically associated with a particular thought or state of mind, and that the two can never be separated. Nonetheless, Descartes has had a huge impact on the science and philosophy that followed. Subjective thoughts and emotion are still seen as less scientific – less amenable to rigorous study, and even less 'real' – than physical, measurable things.

When it comes to medicine, practical advances may have banished the mind even more effectively than philosophical debate. Scholars developed diagnostic tools such as the microscope, stethoscope and blood pressure cuff, and in nineteenth-century Paris, the autopsy. Before that, doctors diagnosed illness based on a patient's account of his or her symptoms; now they could base their conclusions on structural, visible changes. Disease was no longer defined by the subjective experience of the patient, but by the physical condition of the body. It has reached the point where if a patient feels ill but the doctor can't see a problem, it's treated as not being a real disease at all.

Another leap away from subjective experience came in the 1950s with the introduction of randomised controlled trials. To avoid individual biases when testing new therapies, neither doctors nor patients know what treatment is being given, and the results are analysed using rigorous statistical techniques. Unreliable human experience is replaced by hard numbers.

This is arguably one of most important intellectual ideas of modern times. With an objective method of determining which treatments work, doctors are no longer hoodwinked by dodgy cures. Overall, the modern materialist approach to medicine has achieved results that are nothing short of miraculous. We now have antibiotics to banish infection, chemotherapy to fight cancer and vaccines to protect children against killer diseases from polio to measles. We can transplant diseased organs, diagnose Down's syndrome in the womb, and scientists are working on stem cells to repair damaged eyes, hearts and brains.

But this paradigm has been less successful in warding off complex problems such as pain and depression, or stemming the rise of chronic conditions such as heart disease, diabetes and dementia. And it has caused doctors and scientists to discount much about how the body works that to most normal people seems like common sense. The overwhelming focus on the physical – the measurable – has sidelined the more intangible effects of the mind.

That blind spot has allowed the idea of healing thoughts or beliefs to be hijacked by everyone from wishful thinkers to cynical salesmen. Scientific evidence is ignored or grossly distorted. Self-help books, websites and blogs push vastly exaggerated claims: defusing emotional conflict can cure cancer (Ryke Hamer, founder of German New Medicine); our minds can control our DNA (cell biologist Bruce Lipton in his bestselling book *The Biology of Belief*); illness cannot exist in a body that has harmonious thoughts (Rhonda Byrne in the multimillion-selling phenomenon *The Secret*). The mind is marketed as a panacea that can cure our ills without any effort from us save adherence to a rose-tinted world view.

The healing power of the mind – or the lack thereof – has thus become a key battleground in the bigger fight against irrational thinking. The trouble is, the more that sceptics try to debunk wild claims by going on about logic, evidence and the scientific method, the more they isolate those they hope to convert. By denying what seems blatantly obvious to many people – that the mind does influence health; that alternative medicines in many cases do work – they contribute to a lack of trust in, if not a wilful defiance of, science. If scientists say such remedies are worthless, it just proves how much scientists don't know.

What if we take a different approach? By acknowledging the role of the mind in health, can we rescue it from the clutches of pseudoscience?

In writing this book, I travelled around the world to investigate some of the pioneering research that's happening in this area right now. My aim was to track down those scientists swimming against mainstream opinion to study the effects of the mind on the body, and using that knowledge to help patients. What can the mind really do? How does it work, and why? And how can we use these latest findings in our own lives?

We start with perhaps the purest example of mind's influence on the body – the placebo effect – and the scientists looking at what really happens when we take fake pills. After that, we explore some astonishing ways to trick the mind into fighting disease, from using hypnosis to slow gut contractions, to training the immune system to respond to taste and smell. And we learn how simply hearing the right words from your caregiver can determine whether or not you need surgery – and even how long you will live.

The second half of the book moves beyond the immediate effects of thoughts and beliefs to look at how our state of mind shapes disease risk throughout our lives. We visit scientists using brain scanning and DNA analysis to test whether mind–body therapies from meditation to biofeedback really make us healthier. And we look at how our perception of the world around us influences our physical make-up, right down to the activity of our genes.

Along the way, we also come up against the limits of psychological tricks and treatments. What can the mind *not* do? When do the claims made by holistic healers go too far? And what happens when the mind makes things worse?

Writing this book took me further than I ever imagined, from throwing snowballs in a virtual ice canyon to bathing pilgrims in the religious sanctuary of Lourdes. I was inspired by the science I discovered, and by the doctors and researchers fighting resistance at every level – practical, economic and philosophical – to bring the mind and body back together. But most of all I was touched by the patients and trial participants I met, and by their courage and dignity in the face of suffering.

What I learned from them and many others, ultimately, is that the mind is not a panacea. Sometimes it has striking and immediate effects on our bodies. Sometimes it's an important but subtle factor among many, shaping long-term health just as diet or exercise do. Sometimes it has no effect at all. We don't have all the answers yet. But I hope this book will convince the sceptics to reconsider what they might be missing.

And to my friend in the sandpit I would say this: we no longer need to abandon evidence and rational thinking in order to benefit from the curative properties of the mind. The science is there. Let's take a look at what it says.

1

FAKING IT

Why Nothing Works

Until a few months after his second birthday, Parker Beck from Bedford, New Hampshire, seemed to be a happy, healthy little boy. Then he began to withdraw from the world. Parker stopped smiling, speaking or responding to his parents. He woke frequently during the night, made odd, high-pitched screams and developed repetitive habits such as spinning around and banging his head with his hands. After seeking medical advice, his parents Victoria and Gary heard the words they dreaded: their son was showing classic signs of autism. Despite their efforts to gain the best treatment for their son, Parker continued to deteriorate. Until April 1996, that is, when Parker was three. Then something amazing happened.

As is common in children with autism, Parker also had gastro-intestinal problems, including chronic diarrhoea. So Victoria took him to see Karoly Horvath, a gut specialist at the University of Maryland. At Horvath's suggestion, Parker underwent a routine diagnostic test called an endoscopy, in which a camera on the end of a flexible tube is inserted into the intestinal tract. The test itself didn't reveal anything useful. But almost overnight, Parker began to make a dramatic recovery. His gut function improved, and he started sleeping soundly. And he began to communicate again – smiling, making eye contact, and from being almost completely mute, he was suddenly naming flashcards and saying 'Mommy' and 'Daddy' for the first time in over a year.

The label of autism covers a wide spectrum of disorders characterised by problems with language and social interaction, and it affects around half a million children in the United States. Although some children show impaired development from birth, others like Parker appear normal but then regress. Some of the individual symptoms can be treated with drugs. Educational and behavioural therapies (for children and parents) can make a huge difference. But there is no effective treatment or cure. For Victoria, Parker's sudden transformation seemed like a miracle.

She persuaded the hospital to tell her every detail of the endoscopy procedure that Parker had received, right down to the dose of anaesthetic they used. After a process of elimination, she became convinced that the change in her son's symptoms was due to a dose of a gut hormone called secretin. This hormone stimulates the pancreas into producing digestive juices, and was given to Parker as part of a test to make sure that his pancreas was working properly. Victoria believed that there was a connection between her son's gut problems and his symptoms of autism, and concluded that the hormone must have triggered his dramatic improvement.

Desperate to get another dose of secretin for Parker, Victoria called and wrote to the physicians at the University of Maryland to tell them about her theory, but they showed no interest. She also contacted autism researchers and doctors around the country, sending home videos that documented Parker's progress. Finally, in November 1996, her story reached an assistant professor of psychopharmacology at the University of California in Irvine, Kenneth Sokolski, whose son Aaron had autism. Sokolski persuaded a local gastroenterologist to give Aaron the same diagnostic test. He, too, started making eye contact and repeating words.

This was enough to persuade Horvath at the University of Maryland to infuse a third boy with secretin – and he showed the same response. Horvath also gave a second dose to Parker, and Victoria noted another surge in her son's progress. In 1998, Horvath published a report in a medical journal of the three boys' secretin treatment, claiming a 'dramatic improvement in their behavior, manifested by improved eye contact, alertness and expansion of expressive language'.[1]

Horvath refused to give Parker any more doses after that, citing concern that secretin was not licensed for use as a treatment. Victoria eventually found another doctor who was willing to treat Parker, however, and on 7 October 1998, his story was broadcast to an audience of millions on the NBC *Dateline* show. The programme showed the videos of Parker becoming a playful, connected little boy, and featured testimony from other parents who had tried the hormone after hearing of Parker's progress. 'After that secretin, no more diarrhoea, potty trained, looking in the eyes, talking, saying, "Look how pretty outside!"' enthused one mother. 'He was staring right in my face, looking at my eyes, looking like, "Mom, I haven't seen you in a year,"' said another.[2] The *Dateline* programme claimed that of 200 children with autism who had been given the hormone, more than half showed a positive response.

It took just two weeks for Ferring Pharmaceuticals, the only US company licensed to produce secretin, to sell out. Doses of secretin exchanged hands for thousands of dollars on the internet. There were stories of families mortgaging houses to afford it, or buying black-market batches from Mexico and Japan. In the following months, more than 2,500 children were given secretin, and success stories continued to flood in.

'There was tremendous excitement,' recalls paediatrician Adrian Sandler at the Olsen Huff Center for Child Development in Asheville, North Carolina. 'Our phones were ringing off the hook, because parents of kids with autism who we were following wanted to have them treated with secretin.'[3] But medical professionals were concerned about a potential public health crisis. With no hard data on whether secretin was safe to use in repeated doses, let alone whether it worked, more than a dozen clinical trials were urgently commissioned at medical centres across the country. Sandler led the first controlled trial to be published, of 60 autistic children.

As is the gold standard in such trials, Sandler's participants were randomly divided into two groups. One group received the hormone, the other a fake treatment or placebo (in this case, an injection of saline). To be judged an effective drug, secretin would have to do better than placebo. The children's symptoms were assessed before and after the injection by clinicians, parents and teachers who had no idea which treatment each child had received.

Sandler's report appeared in the prestigious *New England Journal of Medicine* in December 1999, and the results were as surprising as they were damning.[4] There was no significant difference between the two groups. The results from the other trials were the same: secretin showed absolutely no benefit when compared to the fake treatment. As a drug for autism, it was useless. The entire promise of secretin was apparently an illusion, invented by parents so desperate to see an improvement in their kids that they had literally imagined it. The secretin story was over.

Or was it? The conclusion in Sandler's paper is one line long: 'A single dose of synthetic human secretin is not an effective treatment for autism.' But what he didn't write in that paper was how struck he was by the fact that *both* groups significantly improved. 'The interesting thing for me was that kids in both groups got better,' he tells me. 'There was a significant treatment response in the group that received secretin and in the group that received saline.'

Was it a lucky coincidence? As with many chronic conditions, symptoms in autism can fluctuate over time. One reason why it is so important to test new treatments against placebo is that any apparent change in symptoms after taking a medicine might be down to chance. But Sandler was surprised by how big the improvement was.

The children in his trial were assessed on an official scale called the Autism Behavior Checklist, which covers a wide range of symptoms from whether they respond to a painful cut or bruise to whether they return a hug. The scale runs from 0 to 158, with higher numbers denoting more severe symptoms. The kids in Sandler's placebo group started the trial with an average score of 63. A month after receiving an injection of the fake hormone (saline solution), they averaged just 45.[5] That's an almost 30% improvement within a few weeks – something that to many parents of kids with autism would seem like a miracle. What's more, the effect was not evenly distributed. Although some children showed no response, others responded dramatically.

This pattern suggested to Sandler that the Becks and other parents convinced of the treatment's benefits had not imagined the changes in their children. Their kids' symptoms really did improve. But it had nothing to do with secretin.

***

Bonnie Anderson didn't notice the water on her kitchen floor until it was too late. One summer evening in 2005, the 75-year-old had fallen asleep on her Davenport sofa while watching TV.[6] She doesn't remember what programme was on, a decorating show, maybe, or an old movie (she doesn't like the bad-language ones, or the bloody kind). When she woke it was dark, and she walked barefoot into the kitchen for a glass of water, without bothering to switch on the lights. But the water purifier had been leaking and she slipped on the wet tiles, landing flat on her back.

Unable to move, Bonnie felt an excruciating pain in her spine. 'It was scary,' she says. 'I thought, "My God, I broke my back."' Her partner, Don, dragged her down the hall and put a blanket over her, and a couple of hours later she was able to get up onto the sofa. Thankfully she wasn't paralysed, but she had fractured her spine – an injury common in elderly people whose bones have been weakened by osteoporosis.

Bonnie lives with Don in a small, white bungalow in Austin, Minnesota. She worked for 40 years as a telephone operator for the town's main employer, Hormel Foods (makers of Spam) and has stayed active into her retirement. She has orange make-up, big white hair and a busy social life, and loves nothing more than an 18-hole round of golf; a sport she has played all her life. But the accident left her devastated. She was in constant pain and couldn't even stand up to do the dishes. 'I couldn't sleep at night,' she says. 'I couldn't play the golf I wanted to play. I'd go and sit in the den with a heating pad.'

A few months later, Bonnie took part in a trial of a promising surgical procedure called vertebroplasty, which injects medical cement into the fractured bone to strengthen it. Don drove Bonnie to the hospital – the Mayo Clinic in Rochester, Minnesota – just before dawn on a cold October morning. She walked out of the hospital after the procedure, and felt better immediately. 'It was wonderful,' she says. 'It really took care of the pain. I was able to go back to my golfing, and everything I wanted to do.'

Almost a decade on, Bonnie is still delighted with the outcome. 'It was a miracle how well it turned out,' she says. Although breathing problems are now starting to slow her down, she isn't limited by her back. 'I have a birthday coming up, I'll be 84,' she chuckles. 'But I still plan on playing a little golf this summer.'

The vertebroplasty apparently healed the effects of Bonnie's fractured spine. Except there's something Bonnie didn't know when she took part in that trial: she wasn't in the vertebroplasty group. The surgery she received was fake.

In 2005, when Bonnie slipped on her wet floor, the technique of vertebroplasty was rapidly gaining popularity. 'Orthopaedic surgeons were doing it. Physiatrists [rehabilitation physicians] were doing it, anaesthesiologists were doing it,' says Jerry Jarvik, a radiologist from the University of Washington in Seattle. 'Anecdotally there were lots and lots of reports as to how effective this procedure was. You'd get them on the procedure table, inject the cement, and they'd effectively jump off cured.'[7]

Bonnie's surgeon at the Mayo Clinic, David Kallmes, says he too had seen 'positive' results from the procedure, with around 80% of his patients getting substantial benefit from it.[8] But nonetheless he was starting to have doubts. The amount of cement that surgeons injected didn't seem to matter much. And Kallmes knew of several cases in which cement was accidentally injected into the wrong part of the spine, and yet the patients still improved. 'There were clues that maybe there was a lot more going on than just the cement,' he says.

To find out what, Kallmes teamed up with Jarvik to do something groundbreaking − at least in the field of surgery. They planned to test the effectiveness of vertebroplasty against a group of patients who would unknowingly receive a pretend operation. Although such placebo-controlled trials are routinely used to test new drugs like secretin, they are not generally required for new surgical procedures, partly because it often isn't seen as ethical to give patients fake surgery. Kallmes points out, however, that with surgery just as with drugs, untested therapies risk harming millions of patients. 'There's nothing unethical about a sham trial or a placebo trial,' he says. 'What is unethical is not doing the trial.'

Kallmes and Jarvik enrolled 131 patients with spinal fractures, including Bonnie, at 11 different medical centres worldwide. Half of them received vertebroplasty and half received a fake procedure. The patients knew that they only had a 50% chance of receiving the cement, but Kallmes went to great lengths to make sure that the sham surgery was as realistic as possible, so that the trial participants wouldn't guess which group they were in. Each patient was taken into the operating room, and a short-acting local anaesthetic was injected into his or her spine. Only then did the surgeon open an envelope to discover whether the patient would receive the real vertebroplasty or not. Either way, the operating team acted out the same predetermined script, saying the same words, opening a tube of the cement so that its characteristic smell of nail polish remover filled the room, and pressing on the patient's back to simulate the placement of the vertebroplasty needles. The only difference was whether or not the surgeon actually injected the cement.

Afterwards, all of the patients were followed for a month, and asked to rate their pain and disability using questionnaires. The study was published in 2009.[9] And even though Kallmes had harboured some doubts about the procedure, he was shocked by the results. Despite all of the apparent benefits of vertebroplasty, there was no significant difference between it and the fake operation.

Both groups substantially improved, however. On average, their pain scores were reduced by almost half, from 7/10 to just 4/10. The disability score was based on a series of questions such as: can you walk a block, or climb stairs without holding a handrail? At the beginning of the trial, the patients answered no to an average of 17 out of 23 questions, a score that is categorised as 'severe disability'. A month after the surgery, they scored on average just 11. Although some were still in pain after the procedure, others, like Bonnie, were practically cured. A second trial of vertebroplasty carried out in Australia was published around the same time, with very similar results.

The patients' improvement was probably due to a range of factors. Pain symptoms can fluctuate, and vertebral fractures do heal, slowly, over time. But both Kallmes and Jarvik believe that to produce such a dramatic improvement, there must have been something else going on – something in the patients' minds. Just as with secretin, it appears that the mere belief they had received a potent treatment was enough to ease – and in some cases banish – their symptoms.

The phenomenon in which people seem to recover after they are given a fake treatment is called the placebo effect, and it is well known in medicine. Clinical trials consistently show a strong placebo effect across a wide range of conditions, from asthma, high blood pressure and gut disorders to morning sickness and erectile dysfunction. In general, however, scientists and doctors view it as a mirage or trick: a statistical anomaly where people would have improved whether they received the treatment or not, combined with a morally dubious phenomenon in which desperate or gullible people are fooled into thinking they are better when they really aren't.

Back in 1954, an article in the medical journal *The Lancet* stated that placebos comfort the ego of 'unintelligent or inadequate patients'.[10] Although doctors might not put it so bluntly today, attitudes haven't changed much since then. The placebo-controlled trials introduced at around that time have been one of the most important developments in medicine, allowing us to determine scientifically which medicines work and which don't, saving count-less lives in the process. They form the bedrock of modern medical practice, and rightly so. But within this framework, the placebo effect is of no interest beyond being something to guard against in clinical trials. If a promising therapy is shown to be no better than placebo, it is thrown out.

Trial results show that neither secretin nor vertebroplasty has any active effect. So according to the rules of evidence-based medicine, the improvements experienced by patients like Parker and Bonnie are worthless.

Yet when Sandler told the parents in his study of secretin that he had found no benefit for the hormone over placebo, a huge 69% of them still wanted it for their kids.[11] Likewise radiologists have refused to give up on vertebroplasty. After Kallmes and Jarvik's report was published, the pair were attacked in hostile editorials and personal letters, and even screamed at in a meeting. 'People felt extraordinarily strongly that we were taking away something that was helping their patients,' says Jarvik. In the US, many insurers still cover the procedure, and even Kallmes still carries out vertebroplasties regardless of his trial results, arguing that for many of his patients there is no other option. 'I see patients get better,' he says. 'So I still do the procedure. You just do what you need to do.'

We see similar cases again and again. In 2012, a popular class of sleeping pills called Z-drugs was shown to be of little value after accounting for the placebo effect.[12] The same year, the sedative ketamine was tested in a double-blind trial for cancer pain; previous studies had described its effects as 'complete', 'dramatic' and 'excellent', yet it too proved to be no better than placebo.[13] In 2014, experts analysed 53 placebo-controlled trials of promising surgical procedures for conditions from angina to arthritic knees, and found that for half of them, sham surgery was just as good.[14]

Perhaps the doctors and patients in all of these cases really were fooled by a combination of random chance and wishful thinking. But by continuing to dismiss the experiences of so many people, I can't help wondering if we are also throwing out something that could be of real help. So here's my question. Might the placebo effect, instead of being an illusion that we should puncture, sometimes be of real clinical value – and if it is, can we harness it *without* exposing patients to potentially risky treatments?

Or to put it another way, can a simple belief – that we are about to get better – have the power to heal?

***

Rosanna Consonni hunches over the desk, gripping its edge with her left hand. In front of her is a grey, rectangular trackpad, and she tentatively places her right index finger on a green circle at its centre. Every few seconds, a red circle lights up at varying positions around the edge of the pad. When that happens, Rosanna has to trace her finger from green to red as quickly as she can.

It's a task that most people would find easy. But the 74-year-old's brow is furrowed in concentration, and she looks like a child struggling to write. She's willing her hand to move but her finger drags slowly, as if it's not really hers. 'Breathe,' advises a young, white-coated neuroscientist, Elisa Frisaldi. Each time Rosanna arrives successfully on red, her time pops up as a blue bar on a graph on Frisaldi's computer screen.

This is the neuroscience department of the Molinette Hospital in Turin, Italy. It is early in the morning and outside the spring sun is shining. A stone's throw away, joggers and dog walkers pass up and down the towpath by the wide, glossy river Po. Blossoms are falling and there are lizards in the grass. But we're squeezed into a windowless basement room packed with computers, lab equipment and a blue couch.

Frisaldi is part of a team headed by one of the pioneers of placebo research, neuroscientist Fabrizio Benedetti. The problem with clinical trials like those of vertebroplasty and secretin is that they are not designed to measure the placebo effect, only to eliminate it. Any changes seen in a placebo group can be due to a range of causes, including random chance, so it's never certain how much improvement, if any, is a result of the placebo itself. Benedetti and Frisaldi, on the other hand, are using carefully controlled laboratory experiments to probe exactly how and when beliefs can ease our symptoms.

Today's volunteer, Rosanna, was 50 when she first noticed that her right hand was trembling. After two years of denial and uncertainty, she finally received a diagnosis: Parkinson's disease. The condition affects about 1 in 500 people; more than half a million in the United States alone. It's a degenerative disease in which brain cells that make a chemical messenger called dopamine gradually die. As levels of dopamine in the brain drop, patients experience steadily worsening symptoms that include stiff muscles, sluggish movement and tremors.

The condition is generally treated with levodopa, a chemical building block that the body converts into dopamine. Rosanna hasn't taken her drug since last night, however, so that her Parkinson's is in full flow for Frisaldi's experiment. She arrives clutching her husband's arm, taking shaky, shuffling steps. Even when she sits, she is in constant motion. She sways as she's speaking, her silver earrings wobbling and her hands waving to and fro. Her chin and throat tremble as if she's chewing. She's wearing kneepads under her grey trousers because she so often falls.

But her spirit appears not to match her frail physical appearance. She is fiercely independent and jokingly refers to her husband, Domenico, as *badente*, or nursemaid. After her initial diagnosis, Rosanna tells me, she didn't want to know anything about her disease. She took her pills, but otherwise 'I didn't read about it. I didn't want to know my future.'[15] For 20 years after her diagnosis, that strategy seemed to work. 'I could drive. I was a good mother. My life didn't change so much.' She enjoyed cycling trips, and snorkelling at the beaches of Versilia, about 150 miles south of Turin.

But in 2008, her symptoms started getting worse. Her body stiffened and her limbs resisted her will to move. One day she went to the supermarket alone, against her doctor's advice, and when a woman in the queue bumped into her she was unable to step to regain her balance. She clattered to the ground and broke her arm. 'I was afraid,' she says. 'I felt something changing in my life.'

Rosanna's doctor recommended surgical intervention, and she now wears a black shoulder strap, attached to a pouch that looks like a small camera bag. It contains a portable infusion pump that delivers her drug continuously, through a plastic tube that dives through her abdomen and into her small intestine. She hates the implant – 'It makes me feel as if I have a handicap,' she says – but it allows her to keep some measure of independence.

Now, with the pump switched off, Frisaldi runs Rosanna through a series of tasks to assess the severity of her symptoms without any drugs. In addition to the track test, she has to circle her arms, walk in a straight line and repeatedly touch her nose. Once the baseline assessment is complete, it's time to open the pouch and activate the pump to begin Rosanna's daily drug infusion. It whirs and beeps; the moment she has been waiting for. 'As soon as I take the drug, I can control my movements better,' she says. 'I feel my hands relaxing, the rigidity in my legs disappearing.' After 45 minutes, I can see what she means. She sits more upright. Her chin is almost still. She moves with more confidence. And her time on the track test is halved.

But how much of this transformation is due to the drug itself, and how much to her expectation of the relief that she is about to feel? This is the type of question that most clinical trials are ill-equipped to address, but that Frisaldi is hoping to answer. Today, Rosanna is getting a full dose of her drug, but on other days she and her fellow volunteers will get a range of different doses, and sometimes they'll know what they're getting and sometimes not (for ethical reasons, Frisaldi isn't allowed to give them no drug at all).

It seems amazing to me that symptoms as severe as Rosanna's – caused by a degenerative neurological disease – might be eased by mere suggestion. But this is what studies of Parkinson's have repeatedly shown. For example, a series of trials carried out by Jon Stoessl, a neurologist at the University of British Columbia in Vancouver, Canada, showed a strong placebo effect when Parkinson's patients were given fake pills.[16] One of them was a keen mountain biker called Paul Pattison. He duly took his capsule and waited for the drug to kick in. 'Boom!' he told the makers of a BBC documentary about the placebo effect.[17] 'My body becomes erect, my shoulders go back.' When he found out he had actually taken a placebo, 'I was in a state of shock. There are physical things that change in me when I take my meds so how could a blank thing, a nothing, create those same feelings?'

Stoessl's experiments answered that question. Using brain scans, he showed that after taking a placebo, the participants' brains were flooded with dopamine, just as when they take their real drug. And it wasn't a small effect – dopamine levels tripled, equivalent to a dose of amphetamine in a healthy person – all from simply *thinking* they had taken their medication.

That finding was followed up by Benedetti, here in Turin. He was carrying out surgery on Parkinson's patients for a therapy called deep brain stimulation. This involves implanting electrodes deep into the brain, in an area called the subthalamic nucleus, which helps to control movement. The neurons in this region are usually kept in check by dopamine, but in Parkinson's patients these cells fire out of control, causing freezing and tremors. Once implanted, the electrodes stimulate these regions and calm the neurons down.

The surgery is done while patients are awake, and Benedetti saw the perfect opportunity to watch the placebo effect in action. The electrode would allow him to monitor activity deep inside the brain as someone takes a placebo – something that isn't usually possible with human volunteers. So he carried out a series of trials: once the electrode was in place, he gave patients a saline injection, and told them it was a powerful anti-Parkinson's drug called apomorphine.

As we wait for Rosanna's drug to kick in, Frisaldi pulls up a series of slides on her computer screen. First, she shows me brain activity that Benedetti recorded before the saline injection. It's a black-and-white line graph, showing the behaviour of a single neuron from the subthalamic nucleus of one of the patients in the study. Each time the neuron fires, the line jumps in a sharp peak. Overall the graph looks like a barcode, a dense forest of spikes that's almost completely black – this is a neuron firing out of control. Then she shows me the activity of the same neuron just after the placebo injection. There's virtual silence; an overwhelmingly white space broken only by the odd, lone spike.

'It's incredible,' says Frisaldi. 'I think it's one of the most impressive studies that Benedetti has done.' Benedetti had chased a belief right down to an individual cell – demonstrating that in Parkinson's patients, motor neurons fire more slowly after injection of a placebo, exactly as they do in response to a real drug.[18]

Between them, what Stoessl and Benedetti showed was remarkable. Although placebo effects had been noted in Parkinson's patients, it never occurred to anyone that placebos might actually mimic the biological effect of treatment. But here was proof that patients weren't imagining their response, or compensating for their symptoms in some other way. The effect was measurable. Real. And physiologically identical to that of the actual drug.

An hour or so later, Rosanna's drug has worn off and the experiment is over. She tells me that she still plans to swim in Versilia this summer, even with her implant, and that she doesn't waste time worrying about how her disease might progress. 'I'm always thinking about the present moment, I don't want to project into the future,' she says. 'That's how I am generally, and the disease hasn't changed that.' She takes out her phone and proudly shows me a picture: 70 kg of lemons from her garden. When she stands to leave, she's tiny and still swaying; she looks like a frail plant being buffeted by the wind.

After learning about the research with Parkinson's patients, I'm impressed by the effect that placebos can have, but I'm left with more questions. If a belief can have the same effect as a drug, why do we need drugs at all? Do placebos work for all conditions, or just some? How does a mere suggestion create a biological effect? To find out, I decide to visit Benedetti himself. But although this is his lab, he's not here. To track him down I have to travel 75 miles north from Turin – and nearly 12,000 feet up.

***

I'm standing on the edge of a cliff, looking down on alpine crows swooping black against the blinding white snow, and across a crinkled blanket of mountain peaks that stretches to the horizon. Sounds are muffled in the thin air, and at −10°C, it's biting cold. Behind me is a huge expanse of ice: the Plateau Rosa glacier. This is 3,500 metres above sea level, on the border between what scientists describe as 'high altitude' and 'very high altitude'. In the Alps, this is almost as high as you can get. From here, only the iconic peak of the Matterhorn rises another kilometre, cutting its crooked triangle out of the azure blue sky.

It is early in the morning, and the plateau is deserted. Then a huge cable car arrives and tips out its load of brightly clad skiers. They pour past me, heading for the shallow slope of the glacier and barely noticing what looks like a metal shed perched on the mountainside. It's half buried in snow and covered in scaffolding.

Inside the shed is Benedetti. He's tall and welcoming, dressed in black ski trousers and a fleece. This is his high-altitude laboratory, packed with equipment and lined with pine slats like a sauna. He shows me round, pointing out the leaking roof – 'It's terrible in summer,' he says – and letting me peek at a three-metre infrared telescope with which he shares this accommodation.

Telescope aside, Benedetti has kitted this space out himself, arranging for all the supplies to be brought in by helicopter. There's a basic living area and kitchen, as well as two bedrooms with bunks, sleep-monitoring equipment and a breathtaking view. The international border runs right through the hut so we step from the living area, which is in Italy, to the lab, which is in Switzerland.

This turns out to be two adjoining rooms, equipped with a mess of machinery and monitors, blinking lights and switches, and bookcases stuffed with files. Wires run across the ceiling and big, green gas canisters lean against the wall. I'm struck by the noise: hums and buzzes, clicks of different frequencies, a periodic hiss. And the thump-thump-thump of an exercise stepper. Working out on the stepper is Benedetti's guinea pig for the day: a stout, young engineer called Davide.

Benedetti is here because the thin air is perfect for studying the placebo effect in another ailment: altitude sickness. Instead of working with ill patients, he can induce symptoms in healthy volunteers simply by bringing them here. Then he plays with their beliefs and expectations, and monitors the physiological effects.

Altitude sickness is caused by a lack of oxygen. As we travel higher above sea level, the percentage of oxygen in the air stays the same, but that air becomes less dense, meaning that there's less oxygen in each lungful that we breathe. Here at 3,500 metres, the oxygen density is only two thirds what it would be at sea level. That can cause symptoms including dizziness, nausea and headaches. The advice to skiers travelling to Plateau Rosa is to allow time to acclimatise by staggering the journey here overnight. To maximise the effects of the altitude for Benedetti's experiment, however, Davide has travelled here in just three hours from sea-level Turin.

With ski poles and a focused expression, Davide looks like an explorer. He's wearing a black neoprene cap fitted with wireless electrodes to monitor his brain activity. Meanwhile various sensors attached to a harness around his chest measure nervous system activity, body and skin temperature, heart activity and the oxygen saturation in his blood. The data are beamed wirelessly from a black recorder, the size of a stopwatch. It's the same 15,000-Euro system that the skydiver Felix Baumgartner used on his record-breaking jump from space,[19] says Benedetti. 'Only we're at 4 kilometres rather than 40 kilometres.'

As Davide works out, Benedetti watches the data come in on his iPad. The engineer's heartbeats are translated into green lines rolling across a black screen, while a digital display shows the oxygen saturation in his blood – at sea level it would normally be around 97–98%, but now it has fallen to just 80%. On a nearby computer screen, a rotating head pulsates with waves of yellow, red and blue – Davide's brain activity.

He steps for 15 minutes, then puts on an oxygen mask attached to a small white canister on his chest, which Benedetti explains will make his activity easier for the remainder of the test. What Benedetti doesn't tell him (or me) is that the mask isn't connected, and the canister is empty. Davide is breathing fake oxygen.

***

I first met Benedetti the evening before, over beer and pizza down in the nearest ski resort of Breuil-Cervinia. Dressed in a zigzag woollen jumper, he looked utterly at home in an alpine lodge. Although he's from the Italian coast, he was always bored on the beach, he tells me. He loves the mountains.

Benedetti sees placebo effects in all aspects of life, from music to sex. He explains that if he gives me a glass of wine and tells me how good it is, that will affect how it tastes to me. Or that if I'm given a hospital room that has a pretty view out of the window, I will recover faster. 'We are symbolic animals,' he says. 'The psychological component is important everywhere.'[20]

His interest in how psychological factors affect our physical bodies began in the 1970s, when he was starting his career as a neuroscientist at the University of Turin. He had already noticed that when he ran clinical trials, patients in the placebo group often did as well as or better than those who received the active drugs. Then he saw a paper that changed his life, not to mention the world's understanding of the placebo effect.

Scientists had recently discovered a class of molecules produced in the brain called endorphins, that act as natural painkillers. Endorphins are opiates, meaning that they belong to the same chemical family as morphine and heroin. The effects that these powerful drugs have on the body were well known, but the fact that we might make our own versions of such molecules was a revelation. It was the first hint that the brain was capable of producing its own drugs.

A neuroscientist called Jon Levine, at the University of California in San Francisco, wondered if this might help to explain how placebos are able to relieve pain. Scientists had generally assumed that gullible patients are somehow tricked into thinking they are in less pain than they actually are. But what if taking a placebo could trigger the release of these natural painkillers? Then the reduction in pain would be real. Levine tested his idea on patients who were in the hospital recovering from oral surgery. Just over a third of them reported significant pain relief after taking a placebo – an intravenous infusion of saline that they thought was a powerful painkiller. Then, without telling them, Levine gave them naloxone, a drug that blocks the effects of endorphins. The patients' pain returned.[21]

It was at this moment, says Benedetti, that 'the biology of placebo was born'. This was the first evidence of biochemical pathways behind the placebo effect. In other words, if someone takes a placebo and feels their pain melt away, it isn't trickery, wishful thinking, or all in the mind. It is a physical mechanism, as concrete as the effects of any drug. Benedetti wondered if this could also explain why the placebo patients in his trials did so well. 'I decided to investigate what was going on in their brains.'

He dedicated his career to lifting the veil of the placebo effect – starting with pain relief. In trials he identified more natural brain chemicals that, triggered by our beliefs, can turn our response to pain up or down. He found that when people take placebo painkillers in place of opioid drugs, these don't just relieve pain, they also slow breathing and heart rate, just as opiates do. And he discovered that some drugs thought to be potent painkillers have no direct effect on pain at all.

Opioid painkillers are supposed to work by binding to endorphin receptors in the brain. This mechanism isn't affected by whether we know we've taken a particular drug. Benedetti showed that in addition to this mode of action, such drugs also work as placebos – they trigger an expectation that our pain will ease, which in turn causes a release of natural endorphins in the brain. This second pathway does depend on us knowing we have taken a drug (and having a positive expectation for it). Incredibly, Benedetti found that some drugs previously thought to be powerful painkillers *only* work in this second way. If you don't know you've taken them, they are useless.

But this is just one placebo mechanism. Benedetti also found pain-relieving placebo effects that are not mediated by endorphins and can't be blocked by naloxone. Then he moved on to studying placebo effects in Parkinson's, the research I learned about from Frisaldi, which works via yet another mechanism: release of dopamine. Placebo effects have only been studied in a few systems so far, but there are probably many others. Benedetti emphasises that the placebo effect isn't a single phenomenon but a 'melting pot' of responses, each using different ingredients from the brain's natural pharmacy.

Up here in the Alps, Benedetti has just started to study how placebos work for altitude sickness. When we're at altitude, low oxygen levels in the blood trigger the brain to produce chemical messengers called prostaglandins. These neurotransmitters cause a variety of physical changes, such as dilating blood vessels, to help pump more oxygen around the body. They are also thought to induce the headaches, dizziness and nausea of altitude sickness. So can fake oxygen interrupt this pathway and ease the symptoms?

Davide finishes his half-hour exercise stint. The altitude has clearly affected him; he looks woozy, and totters slightly as Benedetti helps him to a chair. But he has put in a solid performance on the stepper, impressive for someone who was at sea level just a few hours ago. Benedetti tells me later, after analysing the results from Davide and other volunteers, that the fake oxygen did indeed create a biological effect in their brains compared to a control group who weren't given the placebo. Even though oxygen levels in the blood stayed the same, prostaglandin levels and vasodilation were reduced. When volunteers experience a placebo effect (and not everyone does) their brains respond as if they are breathing real oxygen, reducing their symptoms and allowing them to perform better.

This result illustrates two important points about the limitations of the placebo effect. The first is that any effects caused by belief in a treatment are limited to the natural tools that the body has available. Breathing fake oxygen can cause the brain to respond as if there is more oxygen in the air, but it cannot increase the underlying level of oxygen within the blood. This principle applies to medical conditions too. A placebo might help a patient with cystic fibrosis to breathe a little more easily but it won't create the missing protein that their lungs need, any more than an amputee can grow a new leg. For someone with type 1 diabetes, a placebo can't replace their dose of insulin.

The second point, which is becoming clear from a range of placebo studies, is that effects mediated by expectation tend to be limited to symptoms – things that we are consciously aware of, such as pain, itching, rashes or diarrhoea, as well as cognitive function, sleep and the effects of drugs such as caffeine and alcohol. Placebo effects also seem to be particularly strong for psychiatric disorders such as depression, anxiety and addiction.

In fact, they may be the main mode of action for many psychiatric drugs. Irving Kirsch, a psychologist and associate director of the placebo studies programme at Harvard University, has used freedom-of-information legislation to force the US Food and Drug Administration (FDA) to share clinical trial data sent to it by drug companies. This revealed what the companies had been hiding: that in most cases (severely ill patients are an exception), antidepressant drugs such as Prozac have little effect over and above placebo.[22] Meanwhile Benedetti has found that valium, which is widely prescribed for anxiety disorders, has no effect unless patients know they are taking it.[23] 'The more we know about placebos,' he says, 'the more we learn that many positive outcomes of clinical trials are attributable to placebo effects.'

Placebos, then, are very good at influencing how we *feel*. But there's little evidence that they affect measures we're not consciously aware of, such as cholesterol or blood sugar levels, and they don't seem to address the underlying processes or causes of disease. Bonnie Anderson's fake surgery banished her pain and disability, but it probably didn't mend her spine. One asthma study found that although patients reported that they could breathe more easily after taking a placebo, objective measurements of their lung function did not change.[24] Clinical trials involving cancer patients generally show significant placebo effects for pain and quality of life, but the proportion of patients in placebo groups whose tumours shrink is low (in one analysis of seven trials, it was 2.7%).[25]

These are crucial limitations. Placebos don't create an all-powerful protective magic that can keep us well in every circumstance. We're not going to be able to throw out physical drugs and treatments. But on the other hand, Benedetti's research shows that the effects of placebos are underpinned by measurable, physical changes in the brain and body. And just because the benefits mediated by placebos are mostly subjective, that doesn't mean they have no potential value for medicine.

After all, many of the treatments used in medicine target symptoms rather than underlying disease processes, particularly when the underlying disease is hard to diagnose or treat. Tumour growth and survival time are critical for a cancer patient, but pain control and quality of life are important too. Telling a patient with fibromyalgia or irritable bowel syndrome that there's nothing physically wrong with them will not give them much comfort. A subjective improvement in suicidal thoughts in a patient with depression can mean the difference between life and death.

In lab experiments, placebo effects are often short-lived, but there is evidence that in clinical practice, placebos can keep working for months or years. In a US trial published in 2001, researchers injected neurons from aborted human embryos into the brains of Parkinson's patients, in the hope that they would thrive there and start producing dopamine.[26] The trial was essentially a failure – there was no significant difference between the treatment group and placebo controls. What did make a difference, however, was which group the patients *thought* that they were in. A year later, those who guessed they had received the transplant were doing significantly better (in terms of their own reported scores, and those of blinded medical staff) than those who believed they had received placebo.

Of course, patients who did better might be more likely to guess that they had received the transplant. But the researchers who analysed the data from this study suggest there was more to the effect than that, concluding that even over the course of a year, 'the placebo effect was very strong'.[27] Rosanna believes her refusal to see herself as ill might be one reason why her disease was slow to progress for so many years after her initial diagnosis – this study hints that she might be right.

On the face of it, then, placebos might seem to be a magic pill, with wide-ranging benefits, no side effects, and essentially zero cost. But there has always been one huge problem, which causes even doctors who acknowledge the power of placebos to reject their use in medicine. It has always been assumed that you have to lie to patients for placebos to be effective – to fool them into thinking that they're receiving an active treatment when they are not. No matter what the potential benefit of placebos, critics argue, it is not worth jeopardising the fundamental bond of trust between doctors and patients.

But within the last few years, a handful of scientists have started to suggest that this traditional assumption is wrong. Their results could turn conventional medicine on its head.

2

A DEVIANT IDEA

When Meaning is Everything

Linda Buonanno hugs me as soon as we meet, and shows me upstairs to her small, first-floor apartment in a housing block just off the freeway in Methuen, Massachusetts. Her living space is tidy but densely packed, featuring framed photos, scented candles and the overwhelming colour green. She sits me at the table, in front of a perfectly laid out tea set and a plate of ten macaroons. The 67-year-old is plump with short, auburn hair and a girlish giggle. 'Everyone thinks it's dyed, but it isn't,' she tells me. She hovers until I try a macaroon, then sits down opposite and tells me about her struggles with irritable bowel syndrome (IBS).

She talks fast. Her symptoms first struck two decades ago, when her marriage of 23 years broke down. Although she dreamed of being a hairdresser, she was working shifts in a factory, running machinery that made surgical blades, juggling the 60-hour week with a court battle and caring for the two youngest of her four children. 'I went through hell,' she says. Within a year of the split, she started suffering from intestinal pains, cramps, diarrhoea and bloating.

The condition has affected her ever since, especially at stressful times such as when she was laid off from the factory. Their jobs outsourced to Mexico, the group of women with whom she had worked and bonded was scattered. She retrained as a medical assistant, hoping to find work in a chiropractor's office, but once she qualified she found that no one was hiring. When she did finally find a part-time job, she had to give it up because of the pain from her IBS.

The condition has destroyed her social life too. When the symptoms are bad, 'I can't even leave the house,' she says. 'I'd be keeling over in pain, running to the bathroom all the time.' Even buying groceries requires staying within reach of a loo, and she lists the local facilities: one in the Market Basket, one in the post office down the street. 'This is 20 years I've been doing this,' she says. 'It's a horrible way to live.' Now she has to juggle the condition with looking after her elderly parents – her mother lives alone, while her father, who suffers from dementia, is in a nursing home. Linda's brother was killed in Vietnam, and her twin sister died of cancer 18 years ago, so she is the only one left to help them.

She brightens. 'But I travel,' she says. 'I go to England, I do everything. I love it.' I'm thrown by this statement until I realise that she's talking about Google maps. I ask her to show me, and we move over to her computer, which sits on a desk squeezed between the sofa and the microwave. She fires up the maps programme and lands us on top of Buckingham Palace in London.

Suddenly I get a sense of how much time Linda has spent in this flat. She knows the layout of the palace intimately, zooming in to try and peek through the windows, then flying around the back to check out the private gardens. Other favourite destinations include the Caribbean island of Aruba, and the celebrity mansions of Rodeo Drive. Sometimes she looks up the addresses of her old workmates from the factory, friends who when they lost their jobs moved away to Kentucky or California, places that because of her IBS, and the demands of her parents, she can never visit for real.

Over the years, Linda has, like many patients with irritable bowel syndrome, been passed from doctor to doctor. She has been tested for intolerances and allergies, and has tried cutting out everything from gluten and fat to tomatoes. But she found no relief until she took part in a trial led by Ted Kaptchuk, a professor at Harvard Medical School in Boston. It was a trial that would revolutionise the world of placebo research.

***

'You know I'm deviant?' Ted Kaptchuk looks straight at me and I get the sense that he is rather proud of this fact.[1] 'Yes,' I answer. It's hard to read anything about the Harvard professor without coming across his unusual past. In fact it seeps from every corner of our surroundings – the house where he lives and works, on a leafy side street in Cambridge, Massachusetts.

I'm asked to remove my shoes as I enter, and offered a cup of Earl Grey tea. Persian rugs cover the wooden floors, and proudly displayed in the hall is huge brass tea urn. The décor is elegant, featuring period furniture, modern art and shelves filled with books – rows of hardbound doorstops embossed with gold Chinese lettering next to English volumes from *The Jewish Wardrobe* to *Honey Hunters of Nepal*. Through the window I glimpse the nuanced greens and pinks of a manicured ornamental garden that might be more at home in Japan.

Kaptchuk himself has gold rings, big brown eyes and a sweep of greying hair topped by a black skullcap. He likes to quote from historical manuscripts, and his answers to my questions are accompanied by long pauses and a furrowed brow. I ask him to tell me his own version of the path that brought him here and he says it started when he was a student, and he travelled to Asia to study traditional Chinese medicine.

It's a decision he attributes to 'sixties craziness. I wanted to do something anti-imperialist.' He was also interested in eastern religions and philosophies, and the thinking of the Chinese

Communist leader Mao Zedong. 'Now I think that was a really bad reason to study Chinese medicine. But I didn't wanted to be co-opted, I didn't want to be part of the system.'

After four years in Taiwan and China, he returned to the United States with a degree in Chinese medicine and opened a small acupuncture clinic in Cambridge. He saw patients with all sorts of conditions, mostly chronic complaints from pain to digestive, urinary and respiratory problems. Over the years, however, he became more and more uncomfortable with his role as a healer. He was good at what he did – perhaps too good. He would see dramatic cures, sometimes before patients had even received their treatment. 'I would have patients who left my office totally different,' he says. 'Because they sat and talked to me, and I wrote a prescription. I was petrified that I was psychic. I thought shit, this is crazy.'

Ultimately, Kaptchuk concluded that he didn't have paranormal powers. But equally, he believed that his patients' striking recoveries didn't have anything to do with the needles or the herbs he was prescribing. They were because of something else, and he was interested in finding out what that something was.

In 1998, Harvard Medical School, just down the street from Kaptchuk's clinic, was looking for an expert in Chinese medicine. The US National Institutes of Health (NIH) was opening a centre dedicated to funding scientific research into alternative and complementary medicine. Although tiny compared to existing NIH centres investigating cancer, for example, or genetics, it promised to be a useful new source of research dollars for Harvard. 'But no one there knew a thing about Chinese medicine or any kind of alternative medicine,' says Kaptchuk. 'So they hired me.'

Rather than study Chinese medicine directly, however, he decided to investigate the placebo effect, to find out whether this could explain why his patients did so well. Whereas Benedetti is interested in the molecules and mechanics of the placebo effect, Kaptchuk's focus is on people. The questions he asks are psychological and philosophical. Why should the expectation of a cure affect us so profoundly? Can the placebo effect be split into different components? Is our response affected by factors such as the type of placebo we take, or the bedside manner of our doctor?

In one of his first trials, Kaptchuk compared the effectiveness of two different kinds of placebo – fake acupuncture and a fake pill – in 270 patients with persistent arm pain.[2] It's a study that makes no sense from a conventional perspective. When comparing two inert treatments – nothing with nothing – you wouldn't expect to see any difference. Yet Kaptchuk did see a difference. Placebo acupuncture was more effective for reducing the patients' pain, whereas the placebo pill worked better for helping them to sleep.

This is the problem with placebo effects – in trials they are elusive and ephemeral, rarely disappearing completely but often altering their shape. They change depending on the type of placebo, and they vary in strength between people, conditions and cultures. For example, the percentage of people who responded to placebo in trials of a particular ulcer medication ranged from 59% in Denmark to just 7% in Brazil.[3] The same placebo can have positive, zero or negative effects depending on what we're told about it, and the effects can change over time. Such shifting results have helped to create an aura around the placebo effect, as something slightly unscientific if not downright crazy.

But it isn't crazy. What these results actually show, says Kaptchuk, is that scientists have long got their understanding of the placebo effect backwards. When he arrived at Harvard, he says, the experts there told him that the placebo effect 'was the effect of an inert substance'. It's a commonly used description but one that Kaptchuk describes as 'complete nonsense'. By definition, he points out, an inert substance does not have any effect.

What does have an effect, of course, is our psychological response to those inert substances. Neither fake acupuncture nor a fake pill is in itself capable of doing anything. But patients interpret them in different ways, and that in turn creates different changes in their symptoms.

It's a perspective championed by Dan Moerman, an anthropologist from the University of Michigan, who studied the herbal remedies used by Native American healers before he became interested in placebos – and who analysed those ulcer trials. According to Moerman, the active ingredient is meaning – the meaning that is attached to and surrounds any medical treatment, fake or otherwise. (He wants to change the name of the placebo effect to 'the meaning response', but it's showing no signs of catching on.)

In a phone interview, Moerman refers me to one of Benedetti's studies, on patients recovering from surgery who were given painkillers via an intravenous drip.[4] One group of patients was given the drugs by a doctor who told them what was happening. The other group received their drugs surreptitiously, with the drip controlled by a computer. The only difference between the two groups, says Moerman, was 'human interaction and words'.[5]

The effect of that human interaction was striking. The patients who got their drugs with the doctor present got up to 50% more pain relief. The study included four different drugs, and got the same result for all four. 'I don't see any placebo there at all,' says Moerman. 'What I do see is a clinician wearing a uniform of some sort.' Instead of focusing on fake pills, he argues, we should be looking at those trappings of medicine that make us expect to feel better – whether it's the white coat, stethoscope, and gleaming hospital equipment of a western physician, or the incense and incantations of a traditional healer.

He also points to the clinical trials of antidepressants carried out over the last 30 years. Over that time, drugs have become steadily more effective in treating depression, but so have placebos.[6] Moerman attributes their growing power to media coverage and advertising that has boosted popular awareness and beliefs about the effectiveness of antidepressants. 'Oprah's talking about it, there are ads for antidepressants in every magazine that might be read by someone who is likely to be depressed,' he says. 'Now everybody knows that you can cure depression with a pill.' When we focus on the personal meaning that placebos have for people, rather than on the inert treatments themselves, suddenly the shifting results make perfect sense.

But when Kaptchuk asked patients in clinical trials what they thought about the pills they were taking, he heard something that still didn't quite fit. The central dogma running through all discussions of the placebo effect was that, in order for it to work, you have to believe that you are receiving a real treatment. Patients often experience large placebo effects in trials, where there's a 50/50 chance of being on either the drug or a placebo. Scientists have always assumed, somewhat patronisingly, that this is because people simply forget that they might be on placebo. Yet Kaptchuk found this wasn't the case. 'These people are going crazy on double blind trials,' he says. 'They really worry about whether they are on placebo. They think about it every day.' So how come they were still experiencing placebo effects?

That's when he came up with his boldest – and perhaps most deviant – idea yet.

***

'I was shocked!' says Linda, as I sip my tea and tuck into a second macaroon. She had enrolled in a clinical trial through her gastro-enterologist, Harvard's Anthony Lembo, who was collaborating with Kaptchuk. At the start of the trial, Lembo handed her a bottle of clear plastic capsules with beige powder inside. After so many years of IBS misery, Linda was excited about trying the latest experimental drug for the condition. Then Lembo told her that the pills were placebos, with no active ingredient whatsoever.

Linda knew all about placebos from her training as a medical assistant and thought that taking them was a dumb idea. 'I said,

"Come on, how is a sugar pill going to work?"' she says. 'But I did anything he said, because I was desperate.' She took the bottle home and swallowed the capsules twice a day with a cup of tea.

'I just took it the first day and forgot about it,' she says. Then, something surprising happened. A few days later, she realised that she wasn't sick any more. 'I felt fantastic,' she says. 'No pains, no symptoms, no nothing. I thought, "Wait a minute, this thing works."'

For the three weeks of the trial, Linda went back to living a normal life. She could eat what she wanted, and could go out without worrying about where the nearest bathroom was. She went to the movies with a friend, and for a celebration dinner at the Olive Garden. Then she started to dread the end of the study. 'When it got to the third week I thought, "Oh no, I can't go off these pills."' She begged Lembo to give her more placebos but he explained that he didn't have ethical approval to prescribe them to her once the study was over. Three days after her course of pills ended, her symptoms returned.

Linda wasn't the only patient to benefit from the placebos. Kaptchuk's trial included 80 patients with long-term irritable bowel syndrome, of whom half received a course of placebo. The doctors told these patients that although the capsules contained no active ingredient, they might work through mind–body, self-healing processes.

'Everyone thought it was crazy,' says Kaptchuk. But the trial, published in 2010, found that those patients taking placebos did significantly better than those who received no treatment.[7] Kaptchuk has since got similar results in a pilot study of 20 women with depression,[8] and in a study of 66 migraine patients, who received either drug, placebo or nothing over a total of more than 450 attacks.[9] Taking what they knew to be a placebo reduced their pain by 30% compared with no treatment, says Kaptchuk. 'My team was totally taken aback.'

Linda is now back to square one, but placebo research has been changed forever. One of the big barriers to using placebos in medicine is the concern that it is unethical to deceive patients. Yet Kaptchuk's studies suggest that honest placebos can work, as well.

***

The postman knocks at the door and when I open it he hands me a black cardboard tube marked 'Fragile'. It rattles like a child's toy. Inside, smothered in bubble wrap, is a small, clear plastic jar packed full of blue-and-white capsules, looking just like drugs you might get from the chemist. The label reads 'Metaplacebalin Relaxant Capsules. ONE or TWO capsules to be taken THREE TIMES per day.' My very own placebos.

Since Kaptchuk provided scientific backing for the idea of open-label placebos, it hasn't taken long for a smattering of private companies to start selling them online. A quick Google search turns up Placebo World, Universal Placebos, and Aplacebo, a company based in Chelmsford, UK. Aplacebo's website links to media coverage of Kaptchuk's research and offers a range of products including empty bottles and sprays packaged in different colours for different desired effects (you add your own water), a homeopathic placebo and even a virtual placebo sent by text message.[10]

The products aren't cheap at between £10 and £25, but as the website points out, studies show that the more a placebo costs, the better it works – probably because we instinctively believe that expensive treatments are more effective. When my capsules arrive, I put them in the kitchen cupboard next to the other drugs and they look reassuringly powerful, candy blue, so brightly coloured they almost glow.

A few weeks later, I spend a fraught day looking after two sick children. I finally get them to bed and desperately need the rest of the evening to work, but a throbbing headache is taking hold.

I open the kitchen cupboard and take out the jar. Are Kaptchuk's results a fluke, I wonder? Or can placebos really help us in daily life?

Of course, doctors and drug companies already use placebo effects. As Benedetti's experiment with the open and hidden infusion of painkillers shows, we experience placebo effects every time we receive a drug. Any benefits we ultimately feel are a combination of the active effect of the drug, plus its placebo effect. For some medications, the effects are almost entirely down to their chemical components – placebo statins have little if any effect on cholesterol levels, for example. For others, like antidepressants, it's mostly our minds doing the work.

One approach to harnessing placebo effects, then, is to boost the placebo effect associated with the active drugs we take. One problem with placebos is that they don't work well on everyone (for reasons we'll look at later in this chapter). But there are ways of designing drugs to trigger larger placebo responses in more people. Studies suggest that anything that helps to create the impression of a powerful, potent medication will produce a stronger effect.

Big pills tend to be more effective than small ones, for example. Two pills at once work better than one. A pill with a recognisable brand name stamped across the front is more effective than one without. Coloured pills tend to work better than white ones, although which colour is best depends upon the effect that you are trying to create. Blue tends to help sleep, whereas red is good for relieving pain. Green pills work best for anxiety. The type of intervention matters too: the more dramatic the treatment, the bigger the placebo effect. In general, surgery is better than injections, which are better than capsules, which are better than pills.

There are cultural differences, however, emphasising the point that any effects depend not on placebos themselves, but on what they mean to us. For example, although blue tablets generally make good placebo sleeping pills, they tend to have the opposite effect on Italian men – possibly because blue is the colour of their national football team, so they find it arousing, not relaxing.[11] And although injections make better placebos than pills in the United States, this is not necessarily true in Europe, where there is a stronger cultural belief in the effectiveness of pills.

All fascinating stuff, but can we take the findings on honest placebos to their logical conclusion? Could we knowingly take dud pills to trigger our minds to solve problems such as depression, indigestion, pain or sleep?

Kaptchuk says he loves the idea. 'I certainly think people are over-medicated,' he says. He suggests a good place to start might be conditions where people tend to be on medication long-term, and where the drugs themselves have been shown to have little active effect beyond placebo – such as pain or depression. Then, for those patients who want it, he suggests that they could try a course of placebo first, before progressing if necessary to an active drug.

He doubts whether the idea will catch on with doctors, however. Sometimes in lectures, he says, he asks an audience of physicians whether, given undeniable evidence that honest placebos worked for a particular condition, they would prescribe them. 'Not a single hand goes up.' One such sceptic is Edzard Ernst, professor of alternative medicine at Exeter University, UK, who campaigns against the use of unproven medicines such as homeopathy. He says he is against the idea of using open-label placebos, even if they were shown to help. 'We should always maximise the placebo effect in conjunction with effective treatments,' he explains.[12] Using placebos on their own would mean that patients miss out on the extra therapeutic effect of active drugs.

That certainly makes sense for acute conditions where drugs are proven to be effective. If my son has a serious infection, I want antibiotics for him, not a fake pill. But Kaptchuk argues that in some cases, such as pain, depression or IBS, using placebos on their own might be just as effective as available drugs, and could free people from negative side effects such as addiction. 'I'm hoping that there's going to be some kind of shift because patients want therapies that have fewer side effects,' he says. 'People don't want to go on drugs for long periods of time.'

Ernst counters that there are few conditions for which we have no good treatments at all, and says that where drugs aren't effective there are usually other therapies patients can try (for example physiotherapy or cognitive behavioural therapy). But Kaptchuk's faith in placebos is shared by Simon Bolingbroke, an intelligence analyst from Chelmsford in Essex, and co-founder of Aplacebo, the company that made my capsules.

When I ask Bolingbroke why he decided that trying to sell inert medicines was a good idea, he tells me that he used to be in the military. When serving in Rhodesia (now Zimbabwe) in the 1970s, he was bitten by a tick. After returning home to the UK, he started to suffer from a range of symptoms, including headaches, tiredness and pain in his joints and muscles. His doctors were mystified. By the time his condition was diagnosed as Lyme disease, a bacterial infection spread by ticks, it had spread to his nervous system, damaging it incurably.

Bolingbroke is now in a wheelchair and in constant pain from nerves that fire when they shouldn't. 'It's false pain,' he says. 'My nervous system is not working properly. It's also hard to tell if things are hot or cold. Doing things like cooking or taking a bath, I have to be careful what I touch, because I'm not sure if it's going to burn me.'

He was prescribed multiple medications to deal with the symptoms – at one point he was taking nine different drugs at once, from painkillers to antidepressants. They helped with the pain but he says they started to take over his life and caused dramatic mood swings. 'They were turning me homicidal and suicidal in turns,' he says. 'I wasn't a nice person.'

Inspired by research on placebos, Bolingbroke decided to wean himself off the drugs, slowly replacing them, dose by dose, with inert pills that he made himself. Now, he says, he takes 'virtually no' active medication. When I ask if he is controlling the pain as well on placebos as he was on painkillers, Bolingbroke thinks for a moment, then says, 'It seems apparent to me that I am.'

Now he runs Aplacebo with a friend, selling placebos online. The capsules he sends me are pharmaceutical-grade gelatin casings, the same as conventional medicine in every way except that they are empty. The label is cleverly designed, using jargon to create the impression of a potent, scientific medicine. There's a warning to follow the instruction sheet closely, and the ingredient list looks reassuringly high-tech – nitrogen (78.084%); oxygen (20.946%); argon (0.934%); carbon dioxide (0.039%) – even though all it lists are the chemical components of air.

Despite the persuasive packaging, however, I find it hard to imagine people spending their hard-earned cash on something that openly admits to being nothing. Is Aplacebo really intended as a serious business? 'It sort of started as a joke,' Bolingbroke says. 'We were laughing at ourselves. But it's a joke that's real.' He admits that the company hasn't had any significant sales yet, but insists that with mounting scientific results, and growing awareness of the power of the placebos, his products could one day catch on.

Back in my kitchen, I open the placebo jar and down a couple of capsules with a glass of water, standing by the sink just as I do when taking over-the-counter painkillers. I think about Benedetti's research, picturing his basement lab in Turin, and I try to imagine endorphins flooding my brain. Then I wait to see what happens.

It's hardly a scientific trial, but within 20 minutes or so the pain really does dissipate. With my mini-crisis averted, I can get back to work. And I feel empowered, just a tiny bit, to know that all I needed to do it was my own mind.

***

Bibi Hajerah High School is a ramshackle, mudbrick building in Taluqan, north-eastern Afghanistan. Its female students wear a uniform of black robes and white headscarves, and they take their classes at battered wooden desks lined up in the shade of a tree outside. On the morning of 23 May 2012, classes were progressing as normal when someone complained of a bad smell.

One by one, the girls started to feel sick and dizzy, and to faint. Within hours, more than 100 pupils and teachers had been admitted to the hospital. Pictures broadcast by the media showed armed guards outside the hospital and chaos inside. The crowded wards overflowed with distressed girls, apparently struggling to breathe, being fanned by female relatives.

Khalilullah Aseer, spokesperson for the local police, was sure of the culprit. 'The Afghan people know that the terrorists and the Taliban are doing these things to threaten girls and stop them from going to school,' he told CNN.[13] 'That's something we and the people believe. Now we are implementing democracy in Afghanistan and we want girls to be educated, but the government's enemies don't want this.'

Girls had been strictly forbidden from attending school under the previous Taliban regime, but Afghan women won back the basic right to education when western forces ousted the extremists in 2001. Attending school still took courage, however. Several schoolgirls had suffered acid attacks by the Taliban. Hundreds of girls' schools in Taliban-influenced areas had been closed for safety reasons, and according to one survey, more than half of Afghan parents kept their daughters at home to protect them.

And then, it seemed, there was the poison. The incident at Bibi Hajerah school was the sixth such outbreak in Afghanistan that year. Since 2008, more than 1,600 people from 22 schools across the country had fallen ill in similar circumstances. The poisoning was thought to be a systematic campaign of terror by the Taliban. The Afghan authorities announced several arrests and confessions, and suggested that the victims had succumbed to either toxic gas or a poisoned water supply. Meanwhile local and international media broadcast alarming pictures of victims being carried on stretchers and hooked up to drips.

The symptoms were short-lived, however. The girls all recovered. And although hundreds of samples of blood, urine and water were tested, they came back clear. After interviewing girls and teachers at Bibi Hajerah, staff from the World Health Organization (WHO) concluded that there was no poisoning.[14] The entire outbreak – and probably all of the other episodes too – had been caused by a 'mass psychogenic illness'.

So be warned: the placebo effect has a dark side. The mind might have salutary effects on the body, but it can create negative symptoms too. The official term for this phenomenon is the 'nocebo effect' (Latin for 'I will harm', just as placebo is Latin for 'I will please') and it hasn't been much studied because of ethical concerns. But from what we know about the biology of placebo effects, the Afghan schoolgirls weren't faking it. Fearing or believing that they were about to become ill created real, physical symptoms, even causing some to briefly lose consciousness.

Similar events have been reported throughout history. It may have been mass hysteria that triggered the seventeenth-century witch trials in Salem, Massachusetts. More recently, a fainting epidemic among schoolgirls in the West Bank in 1983 was widely attributed to mass poisoning, with Israel and Palestine blaming each other until official investigators concluded that the symptoms had a psychological cause.

Nocebo effects are even one explanation for the power of voodoo curses. Clifton Meador, a physician at the Vanderbilt School of Medicine in Tennessee, has spent years documenting examples of the nocebo effect. In his book *Symptoms of Unknown Origin* (2005) he tells the story of an Alabama man, 80 years ago, who was cursed with voodoo. By the time the unfortunate patient was seen by a doctor, Drayton Doherty, he was emaciated and apparently close to death. Concluding that nothing he could say would shift the patient's unshakeable belief that he was about to die, Doherty resorted to trickery. With the family's consent, he gave the man a strong emetic then slyly produced a green lizard from his bag, pretending it had come out of the man's body. The witchdoctor had magically hatched the lizard inside him, Doherty told his patient. Now that the evil animal was gone, the man would get well again. And so he did.

It's impossible to confirm Doherty's dramatic account, but these effects aren't only relevant to impressionable schoolgirls or gullible voodoo victims. Anyone can be affected, although who or what can make you feel ill is highly dependent on your social and cultural background, and on what you find believable. If a witch doctor curses you, you might laugh, but if the TV news reports a terrorist gas attack nearby, or a medical doctor in a white coat tells you that you're dying of cancer, you'll be more inclined to take the threat seriously.

Recent studies in the US and UK have induced negative symptoms in volunteers after telling them (falsely) that they were being exposed to powerful wifi radiation, or inhaling environmental toxins.[15] And in 2007, US doctors reported the case of a 29-year-old man from Jackson, Mississippi.[16] He was taking part in a clinical trial for an antidepressant drug and was responding to it well. After an argument with his girlfriend, however, he overdosed on his remaining capsules and collapsed at his local hospital with a racing heart and worryingly low blood pressure. Medical staff gave him 6 litres of intravenous fluids over 4 hours before the message got through from the trial organisers that the patient had been in their placebo group. His symptoms disappeared within 15 minutes.

In fact, most of the side effects we suffer when we take medicines are not due directly to the drugs at all, but to the nocebo effect. In clinical trials for conditions from depression to breast cancer, around a quarter of patients report adverse side effects – most commonly fatigue, headaches and difficulty concentrating

– even when they are taking a placebo. In one study that specifically addressed this phenomenon, Italian researchers followed 96 men who had been prescribed the beta-blocker atenolol for cardio-vascular disease. Some did not know what drug they were taking, whereas others were told about the drug and that it might cause erectile dysfunction. The percentage of patients in each group who subsequently suffered this side effect was 3.1% and 31.2% respectively.[17] This implies that in normal medical practice, where patients know what drug they are taking and are warned of this side effect, as many as a third of them might suffer from impotence after taking atenolol. But only a tenth of those cases are caused by the drug itself. The rest are triggered by the patients' minds.

Although the nocebo effect might seem harmful, from an evolutionary point of view it makes very good sense. Nicholas Humphrey, a theoretical psychologist based in Cambridge, UK, who has written extensively on the evolution of placebo and nocebo effects, argues that if we see other people getting sick around us, or have good reason to believe that we have been poisoned, then to start vomiting is actually a wise strategy.[18] If we really have been poisoned then such early action could be life-saving. If not, then no real harm has been done. Headaches, dizziness and fainting may all serve as warning signals that we should flee a location that could be dangerous, and that we may need medical attention.

From this perspective, the nocebo effect is a biological message that we can't ignore, triggered by psychological cues in our environment that something is wrong. The more threatening we perceive our surroundings to be, the more sensitive we are to such symptoms. But they can be triggered in anyone if the suggestion is strong enough. It's a self-preservation mechanism, or as Kaptchuk puts it, it's what happens 'when you're in a forest that's full of snakes and you see a stick and your brain sees a snake'.

And this, finally, may also explain why we experience positive placebo effects. If threat, anxiety and negative suggestion can induce symptoms of pain and sickness, then it follows that feeling safe and secure, or believing that we are about to feel better, will have the reverse effect. We let our guards down and suppress negative symptoms such as pain. Placebos, then, tap into ancient, evolved neural pathways. Humphrey argues that receiving any kind of medical attention – whether fake, alternative or conventional – helps to persuade these primitive brain circuits that we are loved, safe and getting well, and that there is no further need to feel sick.

Kaptchuk thinks this may be why Linda Buonanno and other participants in his trials experienced placebo effects even though they knew that the pills they were taking were inert. One possibility is that they consciously expected a placebo to help them. But Kaptchuk thinks it runs deeper than that. When Linda took that bottle of capsules from her doctor, Tony Lembo, 'She took Tony home,' he says. 'She took home the care, the concern.'

The fact that some people experience bigger placebo effects than others, and that the same person can experience differing placebo effects at different times, suggests that some people may naturally have a higher threshold for negative symptoms than others, but that the threshold can also slide up and down depending on our circumstances. If we perceive ourselves to be in a forest of snakes – like the Afghan schoolgirls surrounded by Taliban threat, or Linda looking after her kids and working shifts while fighting a messy divorce – the body becomes much more sensitive to biological warning signals such as pain.

If this idea is correct, you'd expect placebos to help eliminate those nocebo effects, by removing our anxiety and pushing our threshold back up. When Linda participated in the placebo trial, 'She was in a forest of concerned people,' says Kaptchuk. 'Her body switched something on that reduced her pain. And she stopped paying attention to her pain as much.'

A cunning experiment carried out by Benedetti at Plateau Rosa, published in 2014, supports the idea that in some cases, placebos work by removing pre-existing nocebo effects.[19] Of 76 students who visited his snow-coated lab, those who had been warned to expect nasty headaches as a side effect of the high altitude suffered more and worse headaches than those who had no idea that this was a risk. Benedetti found that in both groups, the headaches had a biological cause – they were associated with increased levels of prostaglandins, which cause blood vessels to dilate.

It was a nice demonstration of the nocebo effect. In low-oxygen conditions, the brain produces prostaglandins as part of a self-protection mechanism to carry more oxygen around the body. In the students who were worried about headaches, this mechanism was amplified. Their anxiety caused the brain to be more cautious than it would otherwise have been, and to take extra measures to protect itself.

When the students took aspirin, this reduced prostaglandin levels and eased the headaches in both groups. But the most interesting result occurred when they took placebo aspirin. This worked too, but it had a smaller effect than the real aspirin and it only worked in the nocebo group. Benedetti concludes that the placebo was only effective at removing the extra nocebo component of the headaches. It worked by relieving anxiety, which caused the brain to ease off on prostaglandin production.

Benedetti doesn't yet know if this principle will hold for any other types of placebo response. But if it does, this could prove to be 'a new way of looking at placebo', he says. Such placebo effects may not influence underlying disease processes. But they do provide a way to maximise our quality of life, whatever our physical state, and demonstrate that we don't always have to believe the symptoms we feel.

***

'I talk to my pills,' anthropologist Dan Moerman confesses cheerfully. 'I say, "Hey guys, I know you're going to do a terrific job."'[20]

He tells me that he has a painful left knee, and that he uses this technique to boost the effect of his painkillers and get the relief he needs from one pill rather than two.

How we take our drugs, he argues, may be just as important as what they look like. Although there's little research in this area, he and other experts suggest that anything we can do that helps us to attach more significance to a treatment – active or placebo – may boost any beneficial effects that we feel.

In other words, don't throw a pill down your throat absent-mindedly as you're racing for the bus. Instead, create a ritual around it. Harald Walach,[21] a psychologist and philosopher of science from Viadrina European University in Frankfurt, Germany, suggests taking a drug at the same time each day – after a morning bath, in a special room, or with a prayer or silent meditation.[22] Alternatively, Irving Kirsch, a psychologist at the University of Hull, UK, who collaborated with Kaptchuk on his IBS study, suggests using visual imagery. To do this, be as specific as you can about the effect that you would like a particular drug or placebo to have. 'Imagine the improvement,' he tells me.[23]

Or you could ask someone else to dispense a chosen treatment. There's little research on this, but experts including Humphrey and Moerman argue that receiving medical help from others is likely to trigger larger placebo responses than taking care of ourselves, because it creates stronger feelings of safety and security. 'While I think it's a really good thing for me to talk to my pills, it would be much better if my wife did it with me,' says Moerman.

Children are particularly amenable to this type of placebo effect. As any parent knows, kissing a child better, drawing a heart around a skinned knee, rubbing cream on a rash, or calming a cough with a spoonful of honey can have a dramatic impact on pain and other forms of discomfort, even if they contain little or no active medical component.

But it seems to work on adults too. In 2008, Kaptchuk published a trial of 262 patients with IBS.[24] It involved no active treatments, just placebo. One group received no treatment, whereas a second group got fake acupuncture from a polite but cold practitioner who didn't engage in conversation. A third group got the placebo acupuncture from a warm, caring practitioner who sat with them for 45 minutes, listening to their concerns and providing reassurance. Kaptchuk wanted to know how much improvement would be a result of the acupuncture itself, and how much would be the extra-supportive bedside care.

In the no treatment group, 28% of patients said they got 'adequate relief' from their symptoms just from being in the trial. Of those who got placebo acupuncture alone, 44% got adequate relief. In the group that received both acupuncture and empathic care, that figure jumped to 62% – as big an effect as has ever been found for any drug tested for IBS.

For Kaptchuk, this and similar studies highlight what is perhaps the most fundamental lesson from research on placebos: the importance of the doctor–patient encounter. If an empathic healer makes us feel cared for and secure, rather than under threat, this alone can trigger significant biological changes that ease our symptoms. Here was the answer to what was happening in his acupuncture practice, years earlier. When his patients improved before they had even received any treatment, it was their interaction with him that made the difference.

Unfortunately, due to budget and time constraints, as well as an emphasis on drugs and physical treatments, there is increasingly little room for the doctor–patient relationship in western medicine. Doctors might have less than ten minutes with a patient, with the production of a prescription note seen by both sides as more important than having a lengthy, reassuring chat. It's a shift that Kaptchuk blames, ironically, on the introduction of placebo-controlled trials in medicine in the 1950s. 'Before that, doctors knew that care was important for their patients, and that they were an active ingredient,' he says. Now, it's all about the data, and the drugs.

Modern medicine's focus on physical data and objective test measurements has undoubtedly allowed huge advances, but Kaptchuk argues that it has also led to an obsession with molecules and biochemical pathways to the exclusion of how we actually *feel*. 'The only reason people pay attention to placebo [now] is because we've found some neurotransmitters that are involved, and because my team and a lot of other teams are finding great things with neuroimaging,' he says. 'As if patients' experiences are not important.'

Alternative medicine has filled the gap. Therapies such as homeopathy and reiki contain no active ingredient and show no benefit in rigorous clinical trials. They are based on principles that from a scientific point of view are nonsensical – almost certainly they do not work in the way that practitioners claim they do. But with long, personal consultations and empathic care, they are perfectly honed to maximise placebo responses. For that reason they probably do provide real relief, particularly for chronic ailments that conventional medicine is not well equipped to treat.

So even if prescribing honest placebos doesn't catch on, Kaptchuk hopes that his work will trigger a wider debate about the importance of reinstating in western medicine the doctor's role as a healer, so that we can benefit from both personal care *and* scientifically proven treatments, not just one or the other. We need to ask, 'How can we administer drugs so that we make them more effective, and make the side effects less prevalent?' he says.

Clearly, the words physicians use to communicate the benefits and side effects of drugs affect how patients respond. (We'll come back to the importance of language in Chapter 7.) But expectations can be passed to patients in much more subtle ways too. In a classic study carried out in 1985, doctors' beliefs about whether they were prescribing a painkiller or placebo dramatically altered the amount of pain felt by their patients – even though what they told those patients didn't change.[25]

Such indirect placebo effects – dependent on the beliefs and attitudes not of patients but their caregivers – are another reason why placebo effects are seen in children (and even animals).[26] In Sandler's secretin study described in Chapter 1, the parents' positive expectations may have influenced their own behaviour, in turn creating a real improvement in their children's symptoms. Alternative remedies such as amber bracelets for teething pain may soothe a baby by calming the parents' anxiety.

In 2012, Kaptchuk induced both placebo and nocebo effects using pictures of faces flashed up so quickly that patients weren't aware of them[27] – supporting the idea that our experience of symptoms such as pain is easily influenced by subliminal cues. 'Words, gaze, silence, body language, all are important,' says Kaptchuk. Although these aspects of care have often been ignored in medicine, he reckons that placebo studies are now helping to trigger debate about their role.

He's a persuasive speaker, but before I get too carried away he reminds me that there are lots of things positive expectations cannot achieve. 'You're not going to change the underlying physiology [of disease],' he says. 'I don't see that in any of the research.' I guess he's right to emphasise those limits. Feeling great isn't everything. We also want to stay alive, and for many conditions, such as allergies, infections, autoimmune diseases or cancer, that underlying physiology is desperately important.

In cases like these, influencing subjective symptoms isn't enough. So I decide to travel to Germany, where researchers are using the mind to infiltrate the front line of the body's physical fight against disease.

3

PAVLOV'S POWER

How to Train your Immune System

Karl-Heinz Wilbers pops open a small plastic case and takes out four foil blister packs of drugs. Myfortic, tacrolimus . . . these are the names he reads every day, and on which his life now depends. Today there's an extra pill, a chunky white capsule that smells slightly of fish. Before he takes it, he switches on the CD player and cues up 'Help Me', by Johnny Cash. And he pours himself a glass of a bright green liquid that smells strongly of lavender.

Karl-Heinz is a retired psychiatrist from Essen in northern Germany. He's an earnest, academic man with a quiet, almost melancholy demeanour and small, wire-rimmed glasses. Sixteen years ago, his kidneys failed. It's not clear why, he says, although the most common causes are diabetes and high blood pressure. He became one of 80,000 Germans who depend on dialysis, a procedure in which a patient's blood is regularly fed through a tube into a machine and filtered to remove waste products before being passed back into their body.

He was hooked up to the machine for nine hours at a time, four-to-five times a week. Karl-Heinz was lucky, he was able to have the dialysis overnight at home. 'But you can't sleep the whole night,' he says. 'Alarms go off. You have to check the machine, change fluids. You have two big needles in your arm.' He shows me a large scar on the inside of his forearm, where the needles sat in his flesh night after night.

He was alive. He could still walk his dog and was able to paint. But his dependence on the dialysis machine made it impossible to travel, and his chances of surviving to enjoy retirement with his wife and daughter weren't good. The average life expectancy for patients on dialysis is just five years.

After 12 years on dialysis, Karl-Heinz was beating the odds. So when he was finally given the chance of a kidney transplant he said 'Yes', albeit with some trepidation. 'After that, my life was very different,' he says. 'The freedom you get. Being mobile.' He tells me that in the four years since the transplant, he and his wife have visited their daughter in the UK's Lake District, something that would have been impossible on dialysis. They have flown to New York twice, and are planning a trip to the south of England.

But he paid a heavy price. He is no longer tied to the dialysis machine, but to stop his body from rejecting the foreign organ he has to take powerful drugs that suppress his immune system every day for the rest of his life. They put him at risk of life-threatening infections, and he lives with the constant threat of cancer.[1] There are neurological side effects; he gets a painful, burning sensation in his feet. And the toxicity of the drugs puts pressure on his precious kidney. Get the dose too low, and his body could reject it. Too high, and the toxicity could cause the organ to fail.

'Help Me' is one of Karl-Heinz's favourite songs; he has chosen it because it puts him in a calm, thoughtful frame of mind. As he listens to the lyrics, he swallows the chunky capsule and downs his lavender drink. He knows that unlike the rest of the pills in his plastic case, these contain no active drug. He's taking them as part of a pioneering trial to investigate whether this ritual – the drink, the pill, the music – has the power to shape his body's response to his transplant, in this case suppressing it above and beyond the effects of his drugs alone.

The placebos we've looked at so far are based on conscious belief or expectation. You think a pill or injection will have a certain effect, and then it does. Although such fake treatments can create biological changes in the body, they mostly influence subjective symptoms such as pain – affecting how we feel, not our underlying disease. But Karl-Heinz is hoping that his mind will trigger another type of mechanism that can influence basic biological functions, including the immune system.

Proponents say this phenomenon has the potential to slash drug doses for transplant patients like Karl-Heinz, as well as those suffering from allergies, autoimmune disorders and even cancer. But it's far from mainstream medicine, and most immunologists barely acknowledge that it even exists.

***

Imagine taking a plump, yellow lemon from your fruit bowl. Its skin is smooth to touch, glossy and dotted with pores. Now put the lemon on a plate and cut it into quarters. Juice drips down the knife blade onto your fingers and the smell hits you: sharp and sour. You pick up one of the segments and notice how its flesh glistens, the light shining off hundreds of tiny liquid compartments each full to bursting. Then you bite into it, sucking the rush of acid juice onto your tongue.

Did your mouth pucker as you read that paragraph? Did your salivary glands tingle into action, preparing your tongue for the imminent attack of acid? If so, you must have eaten a lemon before, and you have learned the appropriate physiological response. But here's the crucial point. You no longer have to physically eat a lemon to experience these changes. Your body triggers them automatically in response to the sight, the smell – or just the thought – of a lemon, well before you actually taste the juice.

This form of learning, in which a mental cue drives a physical response, is called conditioning. It was famously discovered by a Russian physiologist called Ivan Pavlov in the 1890s. Pavlov was studying how dogs started salivating when he brought them food. Then he noticed that they started to salivate as soon as he entered the room, whether he was carrying food or not. The dogs had learned to associate his presence with being fed. After a while, they responded to him just as they did to their meat.

Pavlov showed that he could train the dogs to associate any stimulus – an electric shock, say, or a light or bell – with dinner. Once the association was learned, that signal on its own was enough to make the dogs drool. It's a beautiful example of how the body doesn't just react blindly to physical events and changes – lemon juice hitting our tongue, for example. It uses psychological cues to stay one step ahead.

Such anticipatory responses prepare us for important biological events such as eating or sex. Your tummy rumbles when you perceive signs – the clock, perhaps, or news headlines on the radio – that tell you it's time for lunch. You get excited by the smell of a lover's perfume, or the sound of their voice. (Psychologists have conditioned volunteers to become sexually aroused by neutral images from guns to penny jars, simply by pairing them with erotic film clips.) The memory of a song your mother used to sing to you at bedtime slows your heart rate and calms you down.

Other conditioned responses have evolved to protect us, preparing us to flee danger, or encouraging us to avoid it. If someone is bitten by a dog in childhood, the sight of a dog later in life may be enough to send their heart racing in fear (this is the basis of many phobias). If we eat a food that gives us a stomach upset, the mere thought or smell of that food may be enough to make us feel sick again. In some cases, even a particular place that we associate with sickness can trigger symptoms. This is why many people having chemotherapy get sick as soon as they arrive at the hospital, before their treatment session even starts.

This much is fairly well known. Pavlov's work on those salivating dogs is world famous. What's less familiar to most scientists, let alone the rest of us, is that conditioning can also trigger placebo responses. If we swallow a pill that contains an active drug, we learn to associate that pill with a particular physiological change. Later, if we receive a look-a-like placebo, we can experience the same change. It's an automatic response in the body that happens regardless of whether we know the pill is fake. But it is triggered via conscious psychological cues – such effects don't occur if we're given a placebo while sedated, for example, or without knowing that we've taken it.

Placebo responses based on physiological conditioning often occur in addition to responses based on conscious expectation. For example, Benedetti tells me that across his trials, the percentage of volunteers who respond to a placebo painkiller is extremely variable, anything from 0% to 100% depending on the circumstances. But if he first gives them a series of identical-looking injections containing an active drug, the proportion who subsequently respond to the placebo soars to a reliable 95–100%. 'You can bet that virtually all patients will respond,' he says – even if they know that the final injection isn't real.[2]

Could such responses be useful in medicine? We heard in Chapter 1 how North Carolina paediatrician Adrian Sandler tested the hormone secretin as a treatment for autism, and found that it was no more effective than placebo. Yet he was struck by how dramatically the children in both groups improved, and he was unable to leave that revelation behind. Any drug that helped as much as the placebo had in his study would be leapt on as a potent treatment. Yet because this remedy involved the mind rather than a pharmaceutical, it was being ignored. In his spare time, Sandler started reading up on placebos, and wondered how he might be able to use them – without deceiving his patients.

The most prevalent diagnosis among the children he saw every day was attention deficit hyperactivity disorder (ADHD). As the name suggests, these kids were inattentive, hyperactive, impulsive.

They were constantly talking and fidgeting, they were unable to wait their turn, and they found it impossible to focus in school. Medication helped them to control their symptoms but still caused problems, from irritable outbursts when the drug wore off in the evening to weight loss and stunted growth. 'It becomes a balancing act in the clinic,' he says, 'trying to find [a dose] that is giving sufficient benefit without giving excessive side effects.'[3]

Sandler wondered whether a placebo might help these children to manage their symptoms on a lower dose of drug. He decided to give his placebos honestly, as part of a regime that would hopefully harness the power of both expectation and conditioning. Seventy ADHD patients, aged six to twelve, completed his two-month trial.

These children were split randomly into three groups. One group underwent a conditioning regime. For one month they received their normal medication, but also swallowed a distinctive green-and-white capsule alongside their drug – they knew this was inert, but Sandler hoped that they would learn to associate it with the physiological response to their active medication. For the second month they received half their usual drug dose, as well as the placebo capsule.

Sandler compared these patients against two control groups, neither of which received any conditioning. One group received their full dose of medication for the first month and a half dose for the second month – just like the conditioning group. The last group got a full dose all the way through.

Sandler published his results in 2010. As expected, in the half-dose control group, the children's symptoms got significantly worse in the second month of the trial. But the conditioned group remained stable, doing just as well as the full-dose patients. In fact there were hints that kids in this group did even better, suffering fewer side effects than those on the full dose of drug.[4]

It is the first and only trial in which honest placebos have been given to children. Sandler says that parents and kids alike embraced the idea, and that more than half of them wanted to keep taking the placebo once the study was over. 'It is the best medicine I've had,' one child told him afterwards. 'I think it tricked the brain into thinking it would work.' Sandler's study is small and preliminary, but combined with Benedetti's findings, it hints that doctors could use simple conditioning procedures to boost the effectiveness of placebos, without any deception required.

For me, that's an exciting finding. By using expectation and conditioning together, ethical placebos could potentially help to reduce drug doses for millions of patients around the world, in conditions from pain and depression to Parkinson's and ADHD.

There's something else about conditioned responses, however, that opens up an entirely new landscape of possibility. These learned, unconscious associations aren't limited to the subjective symptoms – like the distractibility of those with ADHD – that are shaped by conventional placebo effects. They can also influence the immune system, providing a route by which the mind can become a weapon in the body's fight against disease. The mind, in other words, can do much more than help us to feel and perform better. Through conditioning, it might make the difference between life and death.

Scientists denied this was possible until just a few decades ago. Then they were forced to rethink their ideas by two accidental discoveries and a brave teenager called Marette.

***

In 1975, a psychologist called Bob Ader at the University of Rochester in New York was investigating the phenomenon of taste aversion, in which we feel nauseated by a food that has made us sick in the past. He wanted to know how long such learned associations last, so he took a group of rats and fed them several doses of water sweetened with saccharin. This would usually be a treat, but in this experiment he paired the water with injections that made the animals feel sick. Later, Ader gave the rats the sweetened water on its own. Just as he expected, they associated the sugary taste with feeling ill and refused to drink it.

So Ader force-fed it to them using an eyedropper, to see how long it would take for them to forget the negative association. The experiment should have been fairly routine, but what actually happened to the rats seemed like black magic. All Ader gave them during this stage of the experiment was sweetened water, containing no drug whatsoever. But they didn't stop feeling sick. Instead, one by one, they died.[5]

To work out what had killed them, Ader looked more carefully at the chemical he had used to make the rats feel sick in the first place. It was a drug called cytoxan, which as well as causing stomach pain, suppresses the immune system. The dose Ader had used in his experiment was too low to be fatal, so he came to a radical conclusion. When he conditioned the rats, they didn't just learn to feel sick. The extra 'doses' of sweetened water also suppressed their immune systems to the point where they succumbed to fatal infections. It was a stunning finding, suggesting that conditioning reaches far beyond known responses such as salivation, heart rate and blood flow. Our immune systems are vulnerable too.

At the time, this was seen as tantamount to pseudoscience by the immunology establishment. 'The immune system and nervous system were considered to be completely independent systems,' says Manfred Schedlowski, a medical psychologist at the University of Essen in Germany.[6] 'Immunologists thought [Ader's finding] was crazy.' Biologists were convinced that the immune system worked alone, responding to foreign invaders or damage without any help from the brain. Ader died in 2011 but according to his daughter, Deborah, he attributed his insight to the fact that as a psychologist rather than an immunologist, he hadn't been schooled in this dogma. 'I just didn't know any better,' he would say. 'I didn't know that the immune system wasn't supposed to be connected to the brain.'[7]

So although Ader's finding was striking, it wasn't accepted at first. His main problem was that back in the 1970s, he couldn't explain how conditioning of the immune system could possibly work. He was up against generations of immunologists who were convinced that the brain and the immune system don't communicate. They weren't about to change their minds without direct proof of a physical link between the two.

A few years later, they got that proof. David Felten, a neuroscientist working at the Indiana University School of Medicine, was using a powerful microscope to look at body tissues from dissected mice, in order to track where different nerves travelled in the body. In particular, he was interested in the network of the autonomic nervous system, which controls body functions such as heart rate, blood pressure and digestion. Our nerves are divided into the central nervous system, which comprises the brain and spinal cord, and the peripheral nervous system, which runs throughout the body. The peripheral nervous system is in turn divided into two branches. One, the somatic nervous system, deals with conscious messages – it carries our instructions to the muscles so that we can move around and relays sensations such as warmth and pain back to the brain. The second, the autonomic nervous system, controls those physiological systems not usually thought to be under conscious control.

When Felten followed the different branches of the autonomic nervous system, he saw it connecting up with the animals' blood vessels, just as he expected. But then he saw something that seemed totally wrong – nerves running right into the heart of immune organs such as the spleen and thymus (where the body's white blood cells develop and are stored). As he later told a reporter for PBS: 'We saw nerve fibres all over the place, sitting right smack in the middle of some of these cells of the immune system.'[8]

He checked and rechecked his results, making sure that his tissue slices were identified correctly. 'I was almost afraid to say anything. I was worried that we had missed something, and we'd look like a bunch of dufuses.' But there was no escaping the fact that nerves were directly connecting with cells of the immune system. It was incontrovertible evidence of a hard-wired connection between the immune system and the brain.

Felten recalls that when he first published his results in 1981,[9] he was laughed at. But he received encouragement from Jonas Salk, the great US virologist who developed the vaccine that eradicated polio in the 1950s. Felten was so touched by Salk's words that he learned them by heart: 'This research area could turn out to be one of the truly great areas of biology in medicine,' said Salk. 'You'll meet some opposition. Continue to swim upstream.'[10]

Felten started to collaborate with Ader, as well as Ader's colleague Nicholas Cohen, and shortly afterwards he moved to join them at the University of Rochester. These three researchers are now broadly credited with founding a field of research known as 'psychoneuroimmunology'. They championed the idea that in protecting us from illness, the brain and the immune system work together.

Felten's group went on to discover a complex web of connections. As well as hard-wired nerve connections, they found receptors for neurotransmitters – messenger molecules produced by the brain – on the surface of immune cells, as well as new neurotransmitters that could talk to those cells. And they found that the lines of communication went in both directions. Psychological factors such as stress can trigger the release of neurotransmitters that influence immune responses, while chemicals released by the immune system can in turn influence the brain, for example triggering the drowsiness, fever and depressive symptoms that confine us to bed when we are ill.

Meanwhile Ader continued working on conditioned immune responses. The idea of Pavlovian conditioning had soaked into popular culture, but it was usually portrayed as a dubious means for authorities to exert mind control over the masses. In Aldous

Huxley's novel *Brave New World* (1932), toddlers destined for factory work are conditioned to avoid books and flowers using shrill noises and mild electric shocks, while in Anthony Burgess's *A Clockwork Orange* (1962), the protagonist is fed a drug to make him nauseous then forced to watch footage of violent acts. Ader wanted to know if conditioning could instead be harnessed to fight disease.

***

Marette Flies was a cheerful high-school student from Minneapolis, Minnesota. She had a mop of dark, curly hair and a pale, moon-shaped face, and she loved playing the trumpet.

In 1983, when she was 11, she was diagnosed with a life-threatening condition called lupus erythematosus. It's an autoimmune disease, in which the immune system mistakenly attacks the body's own cells. Some autoimmune conditions target a specific organ or cell type: rheumatoid arthritis eats away at the joints, for example, while diabetes kills the cells in the pancreas that make insulin. But with lupus, the immune system wages war on the whole body – the joints, skin, and in severe cases, the heart, kidney, lungs and brain.

Marette was initially treated with steroids to suppress her rampant immune system. She hated taking them – they made her face look like she had 'swallowed a blimp', she complained,[11] and her hair fell out. She'd wake up in the morning with hair covering her pillow. Then she'd eat breakfast, and more hair would fall in her food.

Despite the drug treatment, Marette's condition deteriorated rapidly over the next two years. At first she could still play the trumpet (against her doctors' advice) but then she developed kidney damage, seizures, high blood pressure and bouts of pneumonia. Her immune system also destroyed a vital clotting agent in her blood, causing episodes of severe bleeding. Her condition was so serious that her doctors were considering giving her a hysterectomy, because they feared that when she started to menstruate, she might die from blood loss. Then, in September 1985, her heart began to fail.

With Marette's life in imminent danger, her doctors decided that they had no choice but to put her on a much more powerful immunosuppressant drug. It was cytoxan, the same drug that Ader had used in his experiments on rats. Its use in humans at the time was experimental, and it is highly toxic. The long list of side effects includes vomiting, stomach aches, severe bruising, bleeding, and kidney and liver damage, as well as life-threatening infections and cancer. Cytoxan was Marette's only chance of surviving her lupus, but it was almost as dangerous as the condition itself.

Karen Olness, a paediatrician now at Case Western Reserve University in Ohio, was one of the doctors caring for Marette at the time, using biofeedback and hypnosis to help the teenager cope with the stress and pain of her condition. She had become fond of Marette and was struggling to resign herself to the fact that her patient was not likely to survive this latest crisis. Then Marette's mother, who was a psychologist, showed Olness a copy of one of Ader's papers, which had been published in 1982.[12]

The mice in this latest study had the rodent equivalent of lupus, which could be treated with cytoxan. Ader trained a group of mice to associate cytoxan with saccharin solution, as in his original experiment. Then he kept giving them the sweetened water, along with half of the usual drug dose. Compared to mice that received the reduced drug dose but weren't conditioned, their symptoms were eased and they lived longer, just as in mice given the full dose of the drug. Marette's mother asked Olness if something similar might work in her daughter. Could they train her immune system to respond to a lower dose of the drug, thus sparing her from the worst of the side effects?

Olness called Ader, and he immediately agreed to help design a conditioning regime for Marette. Meanwhile, the ethical committee at Marette's hospital held an emergency meeting to discuss her case. There wasn't a scrap of data from adults or children regarding whether such a trial was safe or would work, the committee noted. This would usually be grounds for immediate rejection. But the danger Marette faced from a full dose of cytoxan was so severe that even though Ader's approach had never been tried in humans before, the committee did something unprecedented. They said yes.

In planning Marette's conditioning regime, Olness's main challenge was to decide what stimulus to pair with the cytoxan. Saccharin worked in the mice because they had never tasted anything sweet before, but it would be too familiar to have much effect in a person. Olness asked Marette what distinctive smells she liked – swimming pools and pot roast, the teenager replied. But those scents don't come in bottles. And to increase the chance of Marette learning a clear association between the stimulus and the drug, Ader told Olness that it should be as peculiar as possible, advising her to choose something strong, unforgettable and previously unknown to Marette.

Olness asked around for suggestions, and tasted vinegars, horehound cough drops, eucalyptus chips and various liqueurs before finally settling on cod liver oil. To this fishy remedy she added a pungent rose perfume, hoping to increase her chances of success by involving Marette's sense of smell as well as her taste buds.

Once the ethical board gave the go-ahead, Marette's treatment started early the next morning. Her doctor placed an intravenous line into Marette's right foot. As the cytoxan infused into her bloodstream, Marette's mother fed her three sips of cod liver oil. The teenager grimaced. 'It makes me feel like vomiting.'[13] Olness uncapped the rose perfume, and wafted the bottle around the room.

Olness repeated this strange ritual – cytoxan, cod liver and rose – once a month for three months. After that, Marette received the cod liver oil and perfume every month, but got the drug only every three months. By the end of the year, she had received just six doses of cytoxan instead of the usual twelve.

Her condition stabilised, and then began to improve.[14] She went for longer periods without being hospitalised, her blood pressure returned to normal, and the clotting factor reappeared in her blood. She had responded exactly as doctors hoped she might on only a fraction of the usual dose of drug. Marette still had lupus, but her symptoms remained under control and she was able to revert to a milder medication. After 15 months she drank no more cod liver oil but she continued to imagine a rose, and was convinced that this thought alone – just as the thought of a lemon can make us salivate – had the power to calm her immune system. She graduated from high school and went to college – driving a sports car and playing trumpet in the college band.

It is impossible to tell from this one case study whether Olness really did succeed in conditioning Marette's immune system or whether her symptoms might have improved anyway. But in 1996, Ader tried a similar approach with ten patients who had multiple sclerosis.[15] He paired their drug, the immunosuppressant cytoxan, with aniseed-flavoured syrup. Subsequently, when given a placebo pill along with the syrup, eight of the patients showed a dampened immune response similar to that usually produced by the active drug. Although only a small trial, it was a further hint that Marette's conditioning really had worked.

Sadly, she didn't live to see it. According to Olness, Marette's heart eventually failed, as a side effect of one of her drugs.[16] She died on Valentine's Day in 1995, aged 22.

***

I'm sitting at a table in the coffee room of the medical psychology department at Essen University Hospital in Germany. I'm joined by two young researchers, Julia Kirchhof and Vanessa Ness. But we're not here for coffee. Kirchhof takes a plastic jug out of the fridge and peels back a layer of clingfilm from the top. Inside is a turquoise-green liquid, so bright it's almost neon. She pours out three glasses and we raise them in a toast. 'It'll stain your teeth and mouth green,' Ness warns me. 'But it doesn't last that long.'

Kirchhof downs hers and frowns. 'Manfred will say it's not strong enough,' she says. It looks strong enough to me, I think, and I take a sip. My eyes are seeing green but immediately I'm hit by a wall of purple, the overwhelming taste of lavender. Otherwise the drink is milky and sweet but also bitter, like drinking bath oil. My mouth screws up, my stomach turns, and my brain does not know what to make of the experience. As the clashing colours combine with the confusion of taste and smell, I can almost feel my neurons firing in bewilderment.

This is the updated version of Olness's cod liver and rose mixture – strawberry milk mixed with green food colouring and a glug of lavender essential oil. It is the invention of medical psychologist Manfred Schedlowski, who is now following up on Ader's intriguing experiments.

After our drink I head to his office, hoping that my teeth aren't embarrassingly green. It's light and spacious, punctuated by red leather armchairs, a black cube coffee table, and a row of geometric art canvases painted by his wife. Schedlowski amiably offers me an armchair and settles himself opposite. He's tall and lanky, with floppy blond hair and a handlebar moustache. When a colleague comes in to warn us that part of the hospital campus is being evacuated, as an unexploded Second World War bomb has just been unearthed on a nearby building site, Schedlowski is unphased. 'Bet it's one of yours!' he says to me cheerily.

Schedlowski has spent the last 15 years trying to turn conditioned immune responses from the intriguing but ultimately anecdotal phenomenon that Ader discovered into a scientifically proven therapy. He started in dramatic fashion by transplanting second hearts into the abdomens of a group of rats. 'It sounds complicated but it's actually a very basic experimental protocol,' he assures me. In rats that received the transplant but were given no medication, the extra heart survived for an average of ten days before it was rejected by the host animal. In rats given several doses of an immunosuppressant drug, the transplant survived three days longer.

Schedlowski then conditioned a third group to associate the drug with a sweet taste, before transplanting the heart. After the transplant, the only medication they received was sweetened water. This group tolerated the extra hearts for an average of 13 days, as long as in the drug-treated rats.[17] Amazingly, Schedlowski had delayed the rejection of the rats' transplanted hearts without drugs, simply by harnessing the rats' minds.

At the time, 'nobody believed us', he says. But he has since repeated the result in a series of other studies. He has shown that surgically removing the nerve to the spleen (the one that Felten discovered) blocks the effect. And that it is possible to boost the effect by combining the conditioning regime with tiny doses of the immunosuppressant drug. On their own, these tiny doses make no difference to how long the transplanted hearts last. But when used along with conditioning, the survival time is dramatically improved. In one study, 20% of the animals kept their second hearts for months, as long as Schedlowski ran the experiment.[18] The sweet taste, together with a tiny amount of drug, protected the graft better than a full dose of medication.

For experiments on humans, Schedlowski developed the bewildering green drink. In trials with healthy volunteers, he has shown that this conditioning can indeed suppress the immune system in people too, and that if combined with tiny drug doses, the effect seems to last long-term; in other words the learned association doesn't fade. Then, in a trial of 62 people with house dust mite allergy, he trained patients to associate the green drink with the effects of the antihistamine drug desloratadine.[19]

A group of patients who received sham conditioning (they

thought they had been conditioned when actually they hadn't) reported that their allergic symptoms had eased. And when they were given a skin prick test, the red wheals that formed were smaller. Conscious expectation – a straightforward placebo effect – had calmed their symptoms. But when Schedlowski measured the underlying immune response, it was unchanged. Only when he added the conditioning was the number of immune cells suppressed as well.[20]

So could Schedlowski repeat the transplant result in humans? 'That's the $1 million question,' he says.

***

To find out, he teamed up with Oliver Witzke, a nephrologist at Essen University Hospital. Witzke tells me that rejection by the host's immune system is a huge problem for kidney transplant patients. About one in ten transplanted kidneys are lost within the first year. Half of those patients die; the other half must go back on dialysis.[21] 'You need to turn down the immune system hugely to keep the graft alive,' he says.[22] He's engaged in the constant balancing act also faced by Wilbers, keeping the drug doses high enough to prevent rejection, without poisoning the kidney that he is trying to save.

He says that Schedlowski's work struck a chord with him, because he knows from experience that psychological factors affect the stability of transplants. 'There is a close interaction between the immune system and the brain,' he says. 'I see in my clinic that patients reject their graft if they have a psychological crisis.'

He says it's a particular risk for young patients, whose lives tend to be more volatile. If they go through a relationship break-up, for example, or lose their job because of their illness, their psychological state can decline. 'If they get into an unstable situation they tend to lose the graft.' This is probably partly due to the fact that patients who are stressed or depressed are less likely to take their meds regularly. 'But I've had a number of patients where I am as sure as I can be as a doctor that they are taking their pills.'

Witzke realised that conditioning might provide a way to suppress the immune system using much lower drug doses, therefore saving his patients from some of the most dangerous side effects, particularly toxicity to the kidney. Together, he and Schedlowski came up with a protocol to test the idea in transplant patients. Initially it was too dangerous to take patients off their drugs, so they designed a pilot study to see if the green drink could suppress the immune system over and above the patients' normal drug regime.

One of the patients in that pilot study was Karl-Heinz. He had to drink the lavender-green concoction alongside his normal drug regime, morning and evening for three days. In the second phase of the study he did the same again, but downed the drink as well as a placebo pill two extra times each day. To make the association with the drug as strong as possible, Schedlowski asked the volunteers to keep the environment constant each time they went through this ritual, swallowing their pills and drink in the same place, and listening to the same music. Karl-Heinz tried the rippling synthesised tones of Jean Michel Jarre's 'Oxygène', before settling on the more heartfelt Johnny Cash.

The extra doses of the green drink did indeed suppress the immune systems of all three patients in the study, including Karl-Heinz, reducing the numbers of all of the immune cell populations that Schedlowski measured by an extra 20–40% (on top of the effect of their drugs). On its own, that's not enough to say that the regime definitely works, but it is promising enough that, as I write this, Schedlowski and Witzke are starting a larger trial, with around 50 patients. If that works, they'll try conditioning while taking the patients off some of their drugs.

Ultimately, Schedlowski believes that the technique could help reduce drug doses for patients with other types of transplants as well as autoimmune diseases such as lupus and multiple sclerosis. And perhaps even cancer. In a series of experiments carried out at the University of Alabama in the 1980s and 1990s, researchers trained mice to associate the flavour of camphor with a drug that activates natural killer cells (a type of immune cell that helps to fight cancer), then transplanted aggressive tumours into their bodies. After the transplant, the conditioned mice received no drugs, just camphor, yet they survived longer than mice given immunotherapy treatment. In one experiment, two of the conditioned animals banished their cancer altogether, despite receiving no active drug.[23] These studies suggested that by boosting the rats' immune systems, conditioning alone had saved their lives.

Using conditioning to reduce drug doses for transplant patients is likely several years away, and for cancer even longer – the Alabama experiments are preliminary, and have never been tried in humans. But Schedlowski says that for less serious conditions, there's no practical reason why doctors couldn't start using conditioning-enhanced therapies straight away.

In one of the last trials that Ader carried out before he died in 2011, for example, psoriasis patients did just as well with conditioning plus a quarter or half dose of corticosteroid ointment as a control group did on the full drug dose.[24] Schedlowski is working with colleagues to design an asthma inhaler that would sometimes dispense a placebo and sometimes the active drug. Sandler's ADHD trial suggests that millions of children could be helped to manage their symptoms on much lower drug doses.

Harnessing conditioned responses to replace drugs with placebos is called Placebo Controlled Dose Reduction (PCDR), and in addition to reducing side effects, it could save billions of dollars in healthcare costs (in 2007, ADHD drugs in the US alone were estimated to cost $5.3 billion).[25]

Unfortunately, scientists are struggling to fund the research they need to get such therapies to the clinic. Sandler says he would love to carry out a larger ADHD trial, but his applications have been rejected. 'I think it's a highly unusual kind of study,' he says. 'The idea of using placebos in open-label to treat a condition is innovative, it turns things upside down. Some reviewers may find that hard to accept.'

And apart from Schedlowski, there is virtually nobody investigating conditioned immune responses. 'I like to say we are the best in the world,' he jokes. 'Because there is nobody else!' Ader and Felten might have won a theoretical victory in proving that the brain and the immune system communicate, but in practice, most immunologists still prefer to ignore the phenomenon.

It's not something that drug companies are interested in, says Schedlowski. 'They don't like the idea of reducing the doses of medications required.' And like Sandler, he has struggled in the past to persuade academic reviewers. A few years ago, he says, he could only publish in niche journals, and he was forced to move from a position in Switzerland back home to Germany because he couldn't get his work funded.

Things are now turning around, however, partly because of Benedetti's work, which has made the entire field of placebo research more acceptable. 'That opened the door and the minds of reviewers that something is going on here,' says Schedlowski. He even changed the name of the phenomenon that he is working on, to try to make it more palatable. 'Before we called it behavioural conditioning of the immune response. Now we call it an immunosuppressive placebo effect.'

In the meantime, however, millions of patients like Karl-Heinz continue to receive drug doses that are much higher than they perhaps need. He lives with the constant fear that he'll lose his kidney, and with it his independence, his ability to travel and quite possibly his life. He describes the idea of reducing his drug dose with conditioning as 'wonderful', and is keen to take part in future trials.

While he waits for further advances, though, he says he has been helped simply by the demonstration that his mind plays a role in protecting his transplant. 'At home I take my drugs with much more awareness,' he says. Thanks to the trial, he now feels like an active player in his own health rather than a passive recipient of drugs, and the side effects of his medication no longer bother him as much. 'Something's happening,' he says. 'Something I can believe in.'

4

FIGHTING FATIGUE

The Ultimate Prison Break

On the morning of 8 May 1978, two men inched through a swirling landscape of mist, wind and snow. Their beards and shaggy 1970s hair were hidden beneath hooded, padded suits – one man in red, one in blue – and they were wearing bulky boots, gloves, and tinted goggles to protect their eyes from the freezing, blinding white. Exhausted and fighting for breath, the pair paused every few steps to lean on their ice axes, panting with mouths wide open and communicating using hand signals because they were too tired to speak. Then they would struggle forward again, barely conscious, limbs failing, aware that they had nothing left but the will to go on.

A few hundred metres above was their goal: the summit of Mount Everest. The 8,848-metre peak – the world's highest – was first conquered by Edmund Hillary and the Sherpa Tenzing Norgay in 1953. But Hillary, and all the others who had scaled the mountain since, relied on canisters of extra oxygen for their climb. Reinhold Messner, a 33-year-old climber from Italy, and his Austrian climbing partner Peter Habeler, were determined to get there without it.

Climbers and doctors alike said they were mad. At such high altitude, the amount of oxygen available in the air to breathe is only a third of that found at sea level. No one knew what would happen to the body in such conditions but it was generally assumed that the pair risked severe brain damage or worse. Physiologists who had studied climbers during a previous expedition led by Hillary in 1960–61 concluded that oxygen levels at the summit were barely enough to keep someone alive if they were resting, let alone attempting an arduous climb.

But Messner was used to facing death in the Himalayas. Eight years earlier he had lost his brother to an avalanche – and seven toes to frostbite – while traversing the notoriously dangerous peak of Nanga Parbat. More recently he had climbed 8068-metre Gasherbrum without oxygen. Whether he made it to the top of Everest or not, he was determined to reach the limit that the human body was capable of.

He and Habeler set off from a camp at 7,985 metres early on the morning of 8 May. As they neared the summit, their progress was increasingly slow. They were forced to climb ridges of rock, as trekking through the deep snow was too exhausting. Breathing was so strenuous that they scarcely had strength for anything else. It became harder and harder to stay on their feet, until eventually they had to collapse into the snow every few steps to rest before crawling on again. They knew that each metre they climbed might be the one that took them past a fatal boundary, the one that ensured they would never return. 'I was attacked by a suffocating fear of death,' wrote Habeler later. 'Now the lack of oxygen is beginning its deadly work.'[1]

Finally, between 1 and 2 p.m., the men saw a metal tripod left behind by Chinese surveyors in 1975. They had reached the summit. Habeler stammered and cried, his tears running from under his goggles into his beard and freezing on his cheeks. Messner says he just sat, legs dangling, with nothing to do at last but breathe: 'I am nothing more than a single, narrow, gasping lung floating over the mists and the summits.'[2]

Messner and Habeler's achievement was a supreme demonstration of endurance, despite bodies and brains that were screaming from lack of oxygen. Yet physiology experiments carried out since then on people exercising at high altitude have revealed a paradox.

It is well known that people reach exhaustion more quickly at altitude. For example, aerobic performance in fit, acclimatised climbers is reduced by around a third at 5,300 metres compared to sea level. The traditional explanation is that in low-oxygen conditions, our blood isn't able to carry as much oxygen around the body. Our muscles become fatigued, and we are unable to continue exercising.

A 2009 study of climbers ascending Everest found that near the summit, at 8,400 metres, the oxygen content of their blood plummeted to just three quarters of normal levels.[3] Messner and Habeler's fear was warranted; if the mountain had been any higher, they may well have not survived. But surprisingly, in blood samples taken at all other altitudes, up to a dizzying 7,100 metres, the oxygen content of the climbers' blood was just the same as recorded at sea level.[4]

In other words, below 7,100 metres, changes in blood oxygen content can't explain climbers' impaired performance at altitude. So what does? It's possible that oxygen diffuses through the tissues less well in such circumstances, suggests study author Daniel Martin, director of the Centre for Altitude, Space and Extreme Environment Medicine at University College London.[5] So even though blood oxygen levels are maintained, less oxygen gets to the cells that need it. But other strange results hint that something else might be going on.

If climbers indeed get tired at altitude because their muscles run out of oxygen, then you'd expect that when exhaustion hits, their hearts should be pumping as fast as possible in an attempt to carry the maximum amount of oxygen around the body. You'd also expect to see particularly high levels in the blood of lactic acid – a toxic waste product that accumulates when the body is short of oxygen. Yet in study after study, scientists have seen neither of these things.[6] People tire at altitude after relatively gentle exercise, even though their hearts still have plenty of reserve capacity. And as they climb higher, the level of lactic acid in their blood at the point of exhaustion actually *falls*.[7]

We fight for breath and struggle to exercise even when oxygen levels in our blood are maintained, with no sign of stress or damage to the brain, muscles or heart.

What is it, then, that slows us down?

***

On 12 August 2012, a 29-year-old Londoner called Mo Farah walked onto the track for arguably the biggest race of his life: the 5,000 metres final at the London Olympic Games. As he approached the starting line, the elated home crowd gave him a standing ovation. A week before, they had seen Farah make history when he won gold in the 10,000 metres. In an event dominated by the African nations of Ethiopia and Kenya, it was the first ever British Olympic win. Now they wanted him to do it again.

But while he had been a strong contender in that race, this was a different proposition. Farah was still recovering from the physical exertion of his victory a week before. And the 5,000 metres was a much greater challenge. He was only eleventh fastest in the world that year, and seven of the faster athletes were lined up next to him, including the quickest of all, the Ethiopian legend Dejen Gebremeskel, who was favourite to win.

Fortunately for Farah, the 12.5-lap race got off to a slow start. He sat back for most of the race, then for the last kilometre fell in second behind Gebremeskel as the pace picked up. In the stands, among thousands of waving union jacks, were his step-daughter and his wife, Tania, heavily pregnant with twins.

Farah pushed to the front, then as the bell rang for the last lap, he opened his stride and slid away from the pack – his slim frame moving easily in white vest and blue shorts, a gold chain bouncing around his neck. Then around the final bend came Gebremeskel, closing fast in green and yellow. It seemed inevitable that the favourite would take the lead, but Farah seemed buoyed by a wave of noise from the crowd. Teeth bared, arms pumping, he somehow pulled away from Gebremeskel and flew over the line first with a look of wide-eyed elation and disbelief.

Farah had run the last mile in four minutes, and his last lap in just 52.94 seconds. BBC commentator Steve Cram (a former distance runner himself) was overcome with emotion. 'My words cannot do justice to how I feel,' he enthused. 'Have you *ever* seen anything like that?'[8] Farah dedicated the two gold medals to his unborn twins.

I was watching the race at home, heavily pregnant myself. Our living room and the whole country were lit up by Farah's achievement. Britain had never won a long distance gold at the Olympics; now we had two. Farah became a national hero. 'The crowd was inspiring,' he said afterwards. 'If it wasn't for them, I don't think I would have dug in as deep.'[9] There seemed no doubt that to win us that medal, Farah had used every scrap of energy, every last muscle fibre, every ounce of will.

So what struck me almost as much as Farah's thrilling sprint finish was what he did right *after* crossing the finish line. Instead of falling to the ground in exhaustion, he started showing off to the crowd with a set of sprightly sit-ups. Then he bounced up again and jogged around the track towards the waiting photographers, arms bent up over his head in his trademark M.

It's a phenomenon that we see often in athletics. World records are broken; sprint finishes won. Athletes apparently marshall every resource to push their bodies to the limit, yet as soon as they cross the finish line, they have the energy and muscle strength to bounce around a victory lap. It raises a similar question to that raised by the climbers on Mount Everest. Why, when it feels as if we're at breaking point, do we have so much left in reserve?

***

Tim Noakes, a sports physiologist at the University of Cape Town, South Africa, is not the type to bow to authority. In fact, he makes a habit of overturning dogmas – sometimes making enemies but also saving athletes' lives.

In the 1980s, for example, he carried out studies that revealed an epidemic of catastrophic neck injuries among rugby players in South Africa.[10] His results were hotly denied at the time but ultimately led to a change in the rules. Then he investigated why so many marathon runners were collapsing. He concluded that it wasn't because of dehydration, as everyone thought, but the reverse: they were drinking too much. According to Noakes, the official advice to runners – that they should drink around 1.5 litres an hour – was poisoning them.

US experts, influenced by the sports drink industry, rejected his findings. The advice wasn't changed until 13% of participants in the 2002 Boston Marathon suffered water intoxication – and one runner died as a result. 'My clash with the multibillion dollar-a-year US sports drink industry taught me that medical science can as easily be bent to serve commercial interest as it can be used to produce "the greatest benefit to humankind",' Noakes said.[11]

Perhaps it's not surprising, then, that Noakes has also spent years attacking one of physiology's most basic assumptions. As an athlete himself, he was interested in fatigue. 'If you are exercising, you are always getting tired and trying to understand why,' he tells me. 'I realised quickly that it wasn't what we were taught.'[12]

The dogma was that athletes get tired when their bodies hit physical limits – their muscles run out of oxygen, fuel or become damaged by the accumulation of toxic byproducts such as lactic acid. This in turn triggers pain and fatigue, forcing us to stop exercising until we recover.

This basic theory had never been questioned since it was proposed by Nobel prizewinning physiologist Archibald Hill in 1923. Yet when Noakes tried to test it, his results didn't make sense. First, Hill's theory predicted that if athletes exercise to their limit, then shortly before they stop with exhaustion oxygen use should level off, because the heart can't pump fast enough to get any more oxygen to the tissues that need it. But just as with the experiments at high altitude, that didn't happen. 'We couldn't find that athletes were running out of oxygen when we tested them,' he says. 'We couldn't see it.'

Meanwhile other studies have shown that although levels of fuel inside muscles (glycogen, fat, ATP) diminish with exercise, they never run out. Noakes also studied muscle use, by asking cyclists to ride exercise bikes with wires attached to their legs. Hill's theory says that athletes should recruit all available resources as they tire, engaging more and more muscle fibres until with nothing more to give, they finally hit breaking point. But Noakes found the reverse. As the cyclists neared exhaustion, muscle fibres were being switched off.[13] At the point at which his volunteers said they felt too fatigued to continue, they were never activating more than about 50% of their available muscle fibres. Exhaustion forced them to stop exercising, yet they had a large reserve of muscle just waiting to be used.

All of this convinced Noakes that the old idea – of fatigue being caused by muscles pushed to the limit – couldn't be true. Instead, he and his colleague, Alan St Clair Gibson, proposed that the feeling of fatigue is imposed centrally, by the brain. Obviously there is a physical limit to what the body can achieve. But rather than responding directly to tired muscles, Noakes and St Clair Gibson proposed that the brain acts in advance of this limit, making us feel tired and forcing us to stop exercising well before any peripheral signs of damage occur. In other words, fatigue isn't a physical event, but a *sensation* or *emotion*, invented by the brain to prevent catastrophic harm. They called the brain system that does this the 'central governor'.[14]

From an evolutionary point of view, such a system makes perfect sense. Relying on signs of damage in the muscles to alert us to fatigue would put us perilously close to collapse every time we exert ourselves. Shutting down physical activity in advance ensures a safe margin of error, and means we can continue to function even after an exhausting challenge. 'We say that's the way humans evolved, because you always need energy afterwards to do other things,' says Noakes. We might suddenly need to run from a predator, for example. 'And when we were hunting, we always had to take the food home.' This is why Farah, even though he ran his heart out to win that second gold, still had energy left for sit-ups and jogging the moment he crossed the finish line.

At altitude, Noakes argues, the effect is even more pronounced. The central governor detects the reduced oxygen in the air and calculates that physical activity in such conditions isn't safe. Even though our muscles are fresh and perfectly capable of exercise, it causes us to feel so fatigued that we can barely walk, and instead diverts our resources to breathing, to make sure that the brain gets enough oxygen. The same thing happens in other potentially threatening environments too. We feel sluggish in hot weather not because our muscles are worn out but because the central governor limits our physical activity in case we overheat. When we are sick, signals from the immune system induce fatigue so that we rest and save our resources to fight the infection.

When Noakes first set out his theory of the central governor a decade or so ago, the notion that the brain – not the heart, lungs or muscles – might ultimately determine the limits of physical performance was seen as ridiculous. Today there's still controversy over his ideas; for example, Everest researcher Martin says that although Noakes 'may well be right' that the central governor, not lack of oxygen, makes us tire so quickly at altitude, this hypothesis is 'not substantiated by any evidence'.[15]

But although exercise physiologists tend to be cautious, psychologists are increasingly convinced that the brain does play an important role in fatigue. For example, many performance-enhancing drugs, such as amphetamines, modafinil and caffeine, work by influencing the central nervous system, not the muscles themselves.[16] Scientists have also stimulated the brain directly with an electric current to boost cyclists' peak power output, and make them feel less tired.[17] Noakes says he hopes that over the next few years, brain-imaging studies will help to prove the existence of the central governor directly.

What intrigues me most about the idea that fatigue is controlled by the brain, however, is whether there is any role for the conscious mind. Can we, in effect, control the central governor?

There's increasing evidence that sometimes we can. A raft of studies has shown that psychological factors can shift our perception of fatigue, adjusting the point at which we feel tired. Athletic performance is influenced for example by our motivation (from monetary reward or the presence of other competitors to the sound of gunshots), whether we are winning or losing, and how far we think we will have to run.

Meanwhile psychologist Chris Beedie at Aberystwyth University in Wales has found that elite cyclists given a pill or drink that they believe is a performance enhancer can cycle on average 2–3% faster[18] – easily enough, in many events, to make the difference between winning a gold medal and failing to place. Beedie suggests this is because the placebo increases their optimism and self-belief, persuading the central governor to free up more resources. 'The brain can do remarkable things but it also limits you,' he says.[19] Taking a placebo lifts those self-imposed constraints. (Placebo expert Fabrizio Benedetti is also a fan of Noakes' ideas, concluding in a paper on fatigue that, 'a placebo may act as a cue signalling the central governor to inhibit its brake'.[20])

So in addition to physical variables such as temperature, oxygen availability, fitness and exertion level, the brain integrates psychological variables such as how confident we are, or how urgent the task is. It then uses the sensation of fatigue to set our maximum pace. If we are anxious about our state of fitness, or uncertain about how far we will need to go, we run slower. But if we are sure about the task ahead, or if we face a life-or-death situation, the governor takes that into account and loosens its grip.

This is why people are capable of feats of physical strength and endurance in emergencies that they would normally find impossible. And if the situation changes, our fatigue level changes too. During a race, we get a sudden boost of energy when we see the finish line. If we're under threat, we feel exhausted as soon as the danger is gone.

When Farah readied for the 5,000 metres, his motivation, confidence and crowd support probably all worked together, persuading his central governor to allow an optimum performance and giving him the edge over his competitors. Meanwhile Messner and Habeler's absolute determination to succeed seems to have pushed them dangerously close to the body's absolute physical limits – to a record-breaking altitude that almost killed them.

The presence of the central governor may explain why interval training – short bursts of high-intensity exercise interspersed by recovery periods – works so well. According to Noakes, regular sprints that push us close to our limit of maximum performance don't just increase physical fitness, they also retrain the brain. They teach the central governor that pushing ourselves that hard was fine, so next time it's safe to push ourselves a little bit further.

But perhaps it is liberating simply to know how over-protective the brain can be. 'You don't have to believe what you are feeling and you don't have to believe what your brain is saying,' says

Noakes. 'However bad you feel, you can carry on and you can still do better.'

***

'It was like being buried alive.' Samantha Miller tells me this matter-of-factly, fixing me with blue eyes as she munches on falafel. 'I was exhausted, with terrible joint pains. It was like having flu all the time with no certainty of recovery. I couldn't do anything. I was trapped.'

Today, Samantha looks vibrant, and younger than her 46 years. She is immaculately dressed in 1950s-inspired floral pinks with a fluffy beret and bright lipstick; her blonde hair is twisted prettily and fixed with a white carnation. We've met for lunch in a Turkish restaurant on London's fashionable Upper Street, and as we talk, she seems energetic, fun and very sharp. It's hard to believe that she has just spent several years fighting her way back from hell on earth.

In the late 1990s, Samantha was living in Hampstead, London, and teaching art at a 'short-staffed, under-funded' secondary school. She found dealing with kids tiring. Children still have 'the invincibility of youth', she says. 'They haven't been crushed by anything yet.' Samantha was also a keen mountain biker and swimmer, and led a hectic social life. If something needed doing, she would pick up the slack. And she was always striving to be perfect.

Then she got ill. 'I had a glandular, viral thing,' she says. It didn't occur to her to take time off work. 'So I was going in with a raging temperature. That was the point at which something changed.' Although she recovered from the illness, afterwards she felt sleepy all the time. A few years later she underwent a back operation, and while she was in the hospital, she contracted gastroenteritis. 'It was horrific,' she says. 'I was being physically assaulted from all sides.'

She recovered from the operation and the gastroenteritis, yet she was left unable even to get out of bed. She was exhausted but not sleeping, in constant pain, and over-sensitive to sound and light. She couldn't get downstairs, so her partner left fruit by the bed when he went to work. She felt overwhelmed and vulnerable – she couldn't sit up, listen to the radio, or answer the door (she remembers reflecting that if instead she was in a wheelchair, having lost the use of her legs completely, she'd at least have had the energy to get to the door).

Whenever she did try to push herself, her symptoms got even worse. So she lay there for months, memorising every crack in the room and staring at a big picture on the wall – an Oxfordshire landscape that she had painted herself. 'I'd think, I can't believe I made that. How can I ever make anything again?'

Although her partner was supportive, she felt that her friends and family didn't understand. They said things like 'I'm exhausted all the time too', and she knew they thought she was somehow choosing to be ill. A particularly painful moment was when her father said, 'This is boring now, I think you should get better.' With no life, and no hope of recovery, Samantha called on her partner and her twin sister. She asked them to help her to kill herself.

***

Chronic fatigue syndrome (CFS) is one of the most controversial conditions in medicine. Researchers, physicians and patients struggle to agree on its name, its definition, or even whether it exists. But the prognosis is bad. A 2005 analysis of trials that followed patients for up to five years concluded that the recovery rate is just 5%.[21]

The condition came to doctors' attention in the twentieth century after a series of mystery epidemics in which large numbers of people were struck by unexplained weakness and fatigue. Two particularly striking outbreaks occurred at the Royal Free Hospital in London in the 1950s, and at Lake Tahoe, Nevada, in the 1980s, where the illness was nicknamed 'Raggedy Ann syndrome'. Then doctors started to see individual cases cropping up in the wider population too.

Chronic fatigue syndrome is also known as myalgic encephalopathy or ME (although not everyone agrees that these are the same condition). There is no proven cause and no agreed diagnostic tests,[22] but the condition is defined as six months or more of persistent fatigue that disrupts life and doesn't get better with rest. It is accompanied by other symptoms including impaired memory or concentration, sore throat, tender lymph nodes, headaches, and joint and muscle pain. In severe cases, like Samantha's, patients are confined to bed for long periods of time.

The symptoms are very similar to those of flu, and in many cases, CFS does seem to be triggered by viral infections such as glandular fever (although not flu itself). The body seems to clear the viral infection, but the fatigue stays. Of adults who get glandular fever, about 12% develop CFS six months later.[23]

Because there is no clear biological mechanism, the condition has often been claimed to have a psychological cause: psychiatrists in the 1970s put it down to 'mass hysteria', while in the 1980s the press cruelly nicknamed it 'yuppie flu', with the implication that sufferers were spoilt young people who were too lazy to work. Medical authorities now agree that it is a genuine discrete condition, even though its causes are still debated, but many sufferers still feel that they are dismissed as hypochondriacs who need to pull themselves together.

Noakes got interested in CFS after seeing athletes affected by the illness, and realised that it did not fit this stereotype. 'I saw too many professional athletes who wanted to run, they were losing everything, and they still couldn't run,' he says. 'The last thing they wanted to do was to be sick.'

He believes the answer to the condition lies in the brain. 'The central governor has got its settings wrong. It's over-estimating how fatigued you are.' Most of the research into the idea of a central governor involves subtle shifts at the very limits of performance, often in elite athletes. But what happens if that entire system crashes? The fatigue that normally protects us from pushing ourselves too far might instead become a prison.

Whatever the trigger – virus, overwork, a genetic predisposition, or (most likely) a combination of several factors – Noakes argues that in CFS the boundaries of physical activity narrow tremendously, to the point where patients are essentially immobilised. If he's right, it would mean that sufferers like Samantha couldn't 'decide' to be more active any more than Messner could have done a jig on top of Everest, or Farah could have shaved 20 seconds off his medal-winning time in London.

But it does hint that their condition might be influenced by psychological factors. Indeed, one of the most robust scientific findings regarding CFS is that when patients are convinced that their condition is biological and untreatable, and fear that engaging in activity will be harmful, they are much less likely to recover. 'If they believe it's incurable, it's incurable,' says Noakes. Although signals from the body are clearly crucial in determining when we tire, ultimately it's the brain that's in control.

This also raises the question of whether cognitive and behavioural therapies could be used to slowly push back the brain's draconian limits. If interval training works for athletes by teaching the central governor that ever-greater levels of exertion are safe, might it also work for patients with CFS?

***

Samantha made a deal with her partner and her sister. She had been referred to a specialist named Peter White at St Bartholomew's Hospital in London. Please, just give him six months, they said. If you're still no better after that, we'll help you to end your life.

Independently of Noakes, White was developing similar ideas about CFS. He doesn't call it a central governor, but he too believes that a combination of triggers – genetic, environmental, psycho-logical – overwhelms the body and throws the nervous system out of balance, causing the brain to reduce massively what it considers a safe level of exertion. To try to reverse the change, he developed with colleagues an approach called graded exercise therapy (GET), which is intended to work like an ultra-gentle form of interval training.

The idea is to set a baseline of activity that the patient can maintain safely, then gradually increase it. Each step has to be small, so as not to risk a relapse. CFS patients report feeling vastly more fatigued than healthy people for a set level of exercise. But White has shown that after a course of GET, they feel less tired after the same amount of exercise, even though their physical fitness is unchanged. Just as when athletes do repeated sprints, the exercise regime slowly retrains the patients' brains that each successive activity level is safe.

White also uses cognitive behavioural therapy (CBT), in which therapists work with patients to challenge negative ideas and beliefs that they have about their illness. This is based on the finding that as long as patients are terrified that any exertion will cause a crash, the fatigue will maintain its vice-like grip. CBT encourages them to try out other ideas and ways of coping, and to test whether small amounts of activity are alright. The hope is that this will reduce their fear, helping them to realise that perhaps some exertion is safe after all and that they have the chance to recover.

White suggested that Samantha try a combination of GET and CBT. 'Will I get better?' Samantha asked her therapist. 'Of course you will,' she replied, and for the first time Samantha believed that it might be true.

Her first exercise goal was simply to turn over in bed once an hour. Every few days, she increased her activity slightly until she was able to sit up for five minutes at a time. Later, when she was out of bed, she might try cooking a meal, but it would be split into parts. Go downstairs. Chop the onions. Go back upstairs and lie down. As a creative person, she found the total lack of spontaneity hard to accept. But the perfectionism that she feels contributed to her condition now helped her.

She kept an activity diary, and as the months progressed she was able to do more. 'Walk two minutes around the block,' she recalls. 'Then walk three minutes. But walking five minutes might put you in bed for three weeks.' She had to stick to the regime, doing no more and no less than the prescribed activity level, no matter how good she was feeling.

If she pushed herself too hard, she would crash. 'It takes incredible discipline,' she says. 'One slip up and you are back to square one.' If she broke the rules and tried to do too much, she would start to feel her body go. 'I'd feel hot from the feet up, almost like I was being poisoned. Then I'd be ruined for weeks.'

It took five years of grim determination, but she finally clawed her way out of the fatigue and back into a normal life.

***

Several small clinical trials suggested that Samantha wasn't alone.[24] The results showed that CBT and GET were helpful treatments. But instead of welcoming the findings, patient groups hated them. 'That was received like a lead balloon by almost all the patient charities in the UK and abroad,' says White.[25] These groups were very sceptical that a 'psychological' treatment like CBT might help patients with CFS, and believed that the activity goals of graded exercise therapy were downright dangerous. CFS is a purely physical condition with no known cure, they argued, so anyone helped by either of White's therapies clearly didn't have it.

Instead, patient groups advocated an approach called pacing. This helps patients adapt to life within the physical limits set by the condition, and encourages them not to do anything that pushes them close to exhaustion. This would make perfect sense if CFS were in fact incurable. But according to White's theories it could be counter-productive, by reinforcing negative beliefs and acting to maintain the condition rather than allowing patients to recover.

Who was right? White and his colleagues decided to do a definitive trial. They worked with the biggest UK patient charity, Action for ME, to design and run the five-year study. It included 641 patients, divided into four groups. A control group just got routine medical care – advice on avoiding extremes of activity, plus drugs for symptoms such as depression, insomnia and pain as needed. The other groups got this standard care plus either CBT, GET or pacing, developed into a therapy (adaptive pacing therapy, or APT).

The researchers published their results in *The Lancet* medical journal in 2011. They found that APT was completely ineffective; patients in this group did no better than the controls.[26] But GET and CBT were both moderately helpful, reducing fatigue and disability scores significantly more than in the other two groups. What's more, 22% of patients recovered after a year in the CBT and GET groups, compared to just 7–8% in the other two groups. That's still not a great success record, but it showed that White's approach was the best treatment available, and demonstrated that recovery from the condition is possible.

If the previous trials had gone down badly, this one was received with absolute fury. *The Lancet* was deluged by letters criticising White's methods. Action for ME rejected the findings. One professor called the trial 'unethical and unscientific' in a 43-page complaint to the journal, while patients used Facebook to ask 'When is the Lancet going to retract this fraudulent study?'

Instead, the journal published an editorial in support of White and his colleagues, saying that they 'should be praised for their willingness to test competing ideas and interventions in a randomised trial'.[27] But it didn't change the attitude of the patient groups. After working for years to fund, organise and run a definitive trial, White finally had the data that he believed could help other CFS patients like Samantha. Patients attending his clinics welcomed the findings, but he could not persuade ME patient organisations to listen.

***

The debate over whether CFS is a biological or a psychological disease still runs hot. In June 2014, two academics from the Essex CFS/ME service at Southend University Hospital, UK, posted an article on the website of the *British Medical Journal*, speculating that CFS might be a 'meme'.[28] This term was invented by geneticist Richard Dawkins in his 1976 book, *The Selfish Gene*, to describe a psychological idea or behaviour that is transmitted from person to person.

The authors of the article argued that several medical conditions through history might be due to memes, such as 'railway brain', a combination of fatigue and psychiatric symptoms that affected travellers on trains in the mid-nineteenth century (a new innovation for the time) and was thought to be due to invisible brain damage caused by the jolty ride. Perhaps, they said, some aspects of CFS are spread in a meme-like fashion too.

There was an immediate campaign to have the article retracted. The ME Association wrote of its members' shock, anger and concern at the suggestions. In online comments beneath the article, CFS patients accused the authors of 'ignorance, bigotry, and outright cruelty', while their ideas were denounced as 'appalling', 'sick and warped' and 'batshit crazy'.[29] A few days later, the Essex CFS/ME service wrote to the ME Association distancing itself from the article and saying that the authors were 'very sorry for any distress they may have caused'.

According to White, the problem, as ever, comes from a mindset that pervades medicine, in which illnesses are seen as either biological or psychological. 'The vast majority of doctors have this dualistic understanding of mind and body,' he says. 'Go and see a psychiatrist for the mind and a physician for the body.' It's a distinction that leaves CFS patients with only two options – their condition is either biological, currently incurable and completely impermeable to psychological factors. Or they're hypochondriacs who have invented the whole thing. No wonder they are on the defensive.

In fact, argues White, it's a false divide. The mind and body inevitably interact and reflect each other: 'What is psychological is physical and what is physical has a psychological perception to it.' Scientists are increasingly finding that psychiatric disorders such as schizophrenia or depression reflect structural abnormalities in the brain, while neurological problems such as Parkinson's cause psychological symptoms as well as physical ones.

White points out that although CBT is often thought of as a psychological therapy, it has physical effects on the body. Several studies have shown that a course of CBT triggers a measurable increase in brain matter, for example, or that it can influence the levels of stress hormones such as cortisol.

A wider shift in attitudes might help CFS patients to accept that physical and psychological factors are entwined in their illness, he argues, without fear of being stigmatised. CFS isn't either biological or psychological. It's both.

***

It is now two years since Samantha recovered from CFS. 'I do more than a lot of women my age,' she says, dipping a strip of pitta bread into her houmous. 'I cycled here. I manage to coordinate my accessories!' She still has to be careful – a challenging bike ride, or getting too stressed at work, can trigger her symptoms. 'You have to step back mentally and physically,' she says.

So now she takes sick days when she's ill, and says, 'No' to things. She works part-time as an art therapist, doing pottery with prison inmates and psychiatric patients with conditions such as bipolar disorder and schizophrenia. Working with clay provides a safe space for them to talk, she says. 'If the conversation gets difficult, you can go right back to the clay.'

She also works as an artist.[30] In one series of pieces, old mementoes – dolls, pinecones, animal skulls – are neatly arranged in ornate frames. She says she likes the idea of rescuing once-precious personal treasures that have become redundant, and giving them a new life and meaning. She paints too, haunting mindscapes including an industrial black and blood-red labyrinth of hospital beds and arched windows, entwined with the first lines of Thomas Hardy's poem, *The Darkling Thrush*: 'I leant upon a coppice gate when frost was spectre-gray. And winter's dregs made desolate the weakening eye of day.'

That poem ends, of course, with the joyous song of a frail thrush; from the dark death of winter, a symbol of 'blessed hope'.

5

IN A TRANCE

Imagine Your Gut as a River

I'm standing in a small hospital room in the north of England. Lying on the bed is a young mother, clutching her abdomen. She's panting and moaning, and she looks terrified.

Emma is 21, with a young son at home. She has blonde hair and a silver charm around her neck. In a chair next to the bed is Emma's own mother. As she strokes her daughter's arm, she fixes the doctor with huge, blue, desperate eyes. She looks as though she hasn't slept in weeks.

Emma is holding a purple hot water bottle against her side; the skin on her arm is red raw from the heat but she refuses to let it go. She groans, moving into different positions in an attempt to ease the pain. She tries sitting on the edge of the bed, then leans over forwards, breathing heavily, with her hand covering her face.

'Owwwww,' she moans, then looks over to apologise. 'Gosh, sorry, it's getting really unbelievable now. It builds.' With the pain, contractions and anxiety, Emma looks just like a woman in labour. Except that there is no baby. And she feels like this every day.

***

We're in the Wythenshawe Hospital in Manchester, UK, and this is just another morning in the clinic of consultant Peter Whorwell. After Emma, he sees Fraser, a man in his late forties who has been diagnosed with congestive myopathy, a heart condition that killed his father in his forties and could now cause Fraser's heart to fail suddenly too.

But that's not why he's here. He can cope with the heart defect, he says – if the worst happens an implanted defibrillator should revive him. What's making him depressed and desperate is permanent, uncontrollable diarrhoea. Fraser shows Whorwell a photograph of a soiled pair of jeans. He wore them to a party, then had to stand with his back against the wall until everyone had gone home.

Then there's 38-year-old Gina, who is visiting the clinic for the first time. 'Tell me about it,' says Whorwell, and Gina talks for maybe half an hour. She started suffering from abdominal pain when she was 18, after giving birth to her daughter. At first, it wasn't clear whether the problem was gastrointestinal or gynaecological. She was given a hysterectomy at 27, and since then several rounds of bowel surgery, but her symptoms have deteriorated each time. Now she's severely constipated. She's taking ten different drugs, including laxatives and powerful painkillers, but nothing solves the problem. Unless she intervenes with lidocaine gel and an anal irrigation system, she doesn't poo for weeks.

She also suffers from searing back and shoulder pain, migraines and stomach aches. She can't sleep because of the pain, and she's exhausted. She works full time, which leaves her with no energy to do anything else, but is determined to support herself rather than relying on benefits. 'I want to show my daughter, you go to work.' Then she calmly asks Whorwell to cut out her colon. 'If a colostomy will fix it, just do it,' she says.

Emma, Fraser and Gina are all suffering from irritable bowel syndrome (IBS), just like Linda Buonanno, who we met in Chapter 2. IBS is frequently dismissed as psychological, and a nuisance rather than life threatening. Yet it's clear from just one morning in Whorwell's clinic that this condition can devastate people's lives.

Around 10–15% of the world's population suffers from the pain, bloating, diarrhoea and constipation of IBS. Conventional treatments are not particularly effective. Doctors offer lifestyle advice (on diet or exercise, for example) or prescribe drugs including laxatives, muscle relaxants and antidepressants, but many patients fail to respond.

Like chronic fatigue syndrome, IBS is a 'functional' disorder, which means that doctors don't see anything physically wrong with the gut in diagnostic tests. And in common with sufferers of chronic fatigue syndrome, IBS patients often feel that they're not taken seriously. 'I'd rather have a broken leg and it would heal in six weeks and I'd be done with it,' says Gina. 'And then people could see, I've got a cast on my leg and that's what's wrong with me. With IBS, people don't understand.'

Whorwell, a world expert in irritable bowel syndrome, thinks the unexplained nature of these conditions probably reflects the inadequacy of medical tests, and that they will eventually be recognised as having a biological basis. But for now, he says, patients are often faced with physicians who use the term 'functional' as a veiled insult, with the implication that they should just pull themselves together: 'Often their doctor has told them it's all in their head.'[1]

He's slim and smartly dressed in trousers and a shirt, with dark brown hair that's flecked with grey. His accent is polished but his sentences are peppered with mild swearwords like 'blooming' and 'bloody' – and 'shit', which once earned him a reprimand when a patient complained. For the most part, though, his patients seem to appreciate his direct manner and sense of humour.

When Whorwell qualified as a gastroenterologist in the 1980s, he was moved by the plight of IBS patients, and felt that they were being let down by the medical profession. Most consultants simply diagnosed them and then discharged them. Instead, Whorwell determined to find a way to help. He had read that hypnosis is a good way to relax muscles, and wondered if it might help to relax the gut, too, so he went on a training course. When he got back, he hypnotised his secretary. 'She almost fell out of her chair,' he says. 'I thought, crikey, this is a potent phenomenon.'

***

Trance states reminiscent of hypnosis have been around for as long as humans have existed, and still exist in traditional cultures around the world. The Kalahari bush people hold ritual healing dances during which painful 'boiling energy' rises up from their bellies. Villagers in Bali change caste by trance dancing with knives and hot coals. Young men in Tibet dance to a drumbeat with needles and spokes in their cheeks, tongues and backs – apparently without feeling pain or bleeding from the wounds. The modern history of hypnosis, however, is generally said to begin in the eighteenth century, with an Austrian doctor called Franz Mesmer; an unfortunate origin that forever set up hypnosis as an enemy of rationalism and science.

Mesmer concocted the idea of a mysterious fluid called 'animal magnetism' that flows through and connects all living things. He claimed that people get sick when this fluid becomes blocked and that he could cure all manner of ills by re-establishing its proper flow. At first he used magnets to manipulate the fluid, but eventually he simply waved his hands to direct it though his patients' bodies – the origin of those melodramatic hand motions that stage hypnotists use today. His patients, afflicted by ailments from paralysis to blindness, typically became more and more agitated before being overcome by convulsions or fainting. When they came round, they pronounced themselves cured.

Working in Paris, he attracted a large and devoted clientele, and mesmerism (the origin of the verb to 'mesmerise') became the height of fashion. Groups of patients (mostly women) sat in big wooden tubs filled with water and iron filings while the doctor walked around them, rubbing his hands over their bodies and triggering hysterical fits.

Paris's conventional medics hated Mesmer for his morally dubious methods, not to mention for stealing their business, and they were desperate to discredit him. In 1784, King Louis XVI convened a panel of top scientists to evaluate Mesmer's technique.[2] Its members included Benjamin Franklin, an expert on the newly discovered force of electricity and America's ambassador to the French court; and Antoine Lavoisier, who discovered oxygen and is often described as the father of modern chemistry.

The king's esteemed panel could detect no trace of Mesmer's magnetic fields using an electrometer and compass. They were also unable to magnetise themselves or members of the public. So Lavoisier designed a series of clever experiments to prove that the effects Mesmer claimed were fake. In one test, a colleague of Mesmer's magnetised a single apricot tree in a garden. A young volunteer was then blindfolded and asked to embrace a series of trees in turn, but was not told which one was magnetised. He became increasingly affected at each tree, before collapsing unconscious by the fourth one. The mesmerist had magnetised only the fifth tree.

'Not a shred of evidence exists for any fluid,' Franklin wrote in his report on Mesmer. 'The practice of magnetisation is the art of increasing the imagination by degrees.'

The commission's cunning investigation provided the model for the clinical trials that form the basis of medicine today. As we saw in Chapter 1, scientists test whether a treatment is effective by comparing it against a fake treatment, or placebo, while subjects are 'blinded' to which one they are receiving just like the young man in the apricot garden. Franklin and his colleagues' work is generally hailed as a pioneering triumph for evidence-based medicine.

But just as controlled trials have caused doctors to overlook the power of the placebo effect, perhaps the king's commission made a similar mistake. They were right to debunk Mesmer's magnetic fluid. But by dismissing his therapy as rubbish because it relied solely on suggestion, did they miss the fact that it really might have the ability to heal?

***

Just let yourself go and relax.

What I notice first are the cards. They're everywhere, 50 or 60 at a rough count, bearing pictures of butterflies, flowers, coastal scenes, dogs in hats. They're covering the desk, lined up on the bookshelves and pinned on the wall. Inside are long, handwritten messages: 'Thank you for everything . . . I just want you to know how grateful I am . . . It has been a huge difference in my life.'

Letting those feelings of relaxation spread through you.

There are posters on the wall too, showing the structure of the gut, and a heavy door painted hospital green, with a notice: 'DO NOT DISTURB. HYPNOTHERAPY IN PROGRESS.' It's quiet except for a ticking clock. The sun slants through the blinds from the car park outside.

Spreading through the little muscles of your feet and ankles. Spreading up to your knees, and to your thighs, and to your tummy.

Most of the space in this office is taken up by two leather armchairs, which face each other. Sitting in the smaller one is Pamela Cruickshanks, a hypnotherapist who has worked with Whorwell at the Wythenshawe Hospital for the last 20 years.

Going to wherever it would be good for it to flow to.

Cruickshanks is leaning forwards, arms folded, notes in her lap. Her eyes are closed. With dark, wiry hair, a necklace made of green square beads and half-rim glasses, she looks like a librarian or a caring aunt. She's speaking softly, in a soothing northern accent that reminds me of caramel.

Imagine that lovely feeling spreading through your shoulders. Down your arms into your hands and fingertips. Through your neck and head, and down through the muscles of your face.

A few feet away, with her feet up in an oversize reclining armchair, is Nicole. Her eyes are closed too, and she's breathing deeply, slowly. The 48-year-old is slim, with chestnut brown hair, silver stud earrings and a slick of lip gloss.

Everything is more comfortable and relaxed. Enjoy that lovely flow.

Fourteen years ago, Nicole was enjoying her career as an air hostess, and expecting her first child. But her son was born with an unexpected cleft lip and palate, as well as hearing and speech problems. Then the father left, taking all of the couple's money. Unable to pay the rent, she became homeless.

Imagine that you are really relaxed before eating. Enjoy your food. Eating slowly, chewing thoroughly, feeling comfortable with your food as it goes down into your stomach.

Within a few weeks, Nicole was suddenly a single mother, with no job, money, partner or home, and a child with special needs. She managed to get herself and her son into a council house, and saw her son through several operations while studying to be a dental nurse. She would get up at five in the morning to study while he slept, before taking him to nursery and going to work.

Your stomach sends little waves through, like waves on the sea on a lovely calm beach. Imagine your gut is learning from the waves of the sea.

But the stress caught up with her. She felt sick and in constant pain, as if razor blades were being dragged inside her body. And she became hugely bloated. Naturally slender, she now looked 9 months pregnant.

Imagine the little waves in the small intestine, moving the food along. Absorbing it into the body.

It took 12 years for Nicole to get a diagnosis of IBS. Her consultant prescribed more and more drugs, until she didn't know which drug was doing what. But nothing helped the pain, vomiting or constant diarrhoea. A low point was when she was admitted to hospital, struggling to breathe, with blood pressure so high that she required emergency treatment and her stomach so distended that staff refused to believe she wasn't pregnant.

Everything is calm and comfortable. See how the water sparkles in the sunshine.

Nicole was transferred to Peter Whorwell's care, and when he suggested that hypnotherapy might help, she was sceptical to say the least. But she was so desperate, she was willing to try anything. Today is her sixth session with Cruickshanks. The lines are gone from her face. She looks serene.

Rather than your tummy controlling you, you control your tummy. I'm asking your unconscious mind to please help. Please put in place the right way for the gut to work.

When Cruickshanks finishes, Nicole takes a deep breath. She scratches, stretches her arms above her head and opens her eyes.

***

Mesmerism didn't disappear after being discredited by the French king's commission. Instead, it was reinvented – and given a new name.

Despite Franklin's scathing report, mesmerists continued to practise into the nineteenth century throughout Europe and the US. Rather than experiencing hysterical convulsions, however, their patients tended to fall into sleep-like trances. This was demonstrated in compelling stage shows, during which practitioners often claimed that the trance state induced paranormal powers such as telepathy and clairvoyance. Not surprisingly, the medical establishment remained convinced that the whole thing was fake.

In 1841, a Scottish doctor called James Braid attended one of these shows intending to debunk it, but after examining the mesmerised subjects, he came away convinced that beneath all the dramatics was something worth studying. He concluded that no hand waving was needed; he could induce a trance in people simply by asking them to focus their attention on an object such as a bottle top or a candle flame. There was nothing paranormal, just a physical phenomenon that could be studied scientifically. He called it 'neurohypnosis' after Hypnos, the Greek god of sleep.

Hypnosis was subsequently embraced by psychotherapists like Sigmund Freud, who used it early in his career to uncover and resolve psychiatric problems, and Milton Erickson, who broke away from the authoritarian approach of earlier hypnotists. Instead he developed indirect methods of suggestion to overcome patients' resistance to being hypnotised, and repeated true phrases during inductions – such as 'You are sitting comfortably' – to gain patients' trust. Both were convinced that the unconscious mind plays an important role in physical health.

For the most part, though, the medical profession has remained unimpressed. Links to wacky practices such as past-life regression; cases in which therapists have unwittingly planted false memories of abuse; and the continuing popularity of stage shows have all added to the reputation of hypnosis as seedy and unscientific.

Another problem is that scientists struggle to understand what hypnosis does to the brain. It turns out that being hypnotised is easy enough to describe, but much harder to explain. 'It's like you're entering the imagined world,' says David Spiegel, a psychiatrist at Stanford University and one of the world's leading researchers in hypnotherapy. 'There's less judgment, there's less contrasting and comparing, you're just in the flow of the experience. What you are experiencing seems very vivid and real. You're not conflicted about doing it, you're not counting the seconds. It's a mental rollercoaster ride where you are just hanging on and seeing what's happening.'[3]

Psychologists usually give a bland definition such as 'a state of highly focused attention combined with suspension of peripheral awareness'. When hypnotised, people seem more suggestible than normal, and more susceptible to distortions of reality such as false memories, amnesia and hallucinations. They can also feel as if they lose conscious control of their actions. For example, if the hypnotist suggests that their arm will raise, it appears to them as if it moves on its own.[4]

A common explanation for these strange effects is that during hypnosis, different parts of our awareness become separated from each other. This means that our unconscious brain can comply with suggestions without our conscious self knowing. The hypnotist asks us to raise our arm and we do so, but it feels as though someone else is lifting it for us. When we experience amnesia, the unconscious mind registers events that occur but these sensations don't make it through to our conscious awareness.

We probably flit in and out of hypnotic states all the time. Have you ever driven from one place to another, and realised when you arrived that you couldn't remember anything about your journey? Or been so caught up in the story of a fascinating book or film that you failed to notice when someone called your name?

This could mean that there's nothing much special going on at all. Indeed, some researchers argue that there isn't any such thing as hypnosis, and that the feats people perform when apparently hypnotised have other explanations, from peer pressure and playacting to having a vivid imagination. Or perhaps it is just a way to boost our expectation that a particular thing is going to happen, like a turbo-charged placebo effect. This nicely explains why hypnosis takes so many forms, from hysterical fits to sleepy stupors to the boiling energy of the Kalahari bush people. Hypnosis is simply a self-fulfilling prophecy, in which whatever people expect to experience comes true.

Recent brain-scanning studies suggest, however, that something significant does happen in the brain when we are hypnotised. One example is what Spiegel calls his 'believing is seeing' experiment.[5] He showed volunteers a series of grids – some colour, some grey-scale – while scanning their brains. Then (while they were still looking at the grids) he told them that the colour grid was black and white, and that the black-and-white grid was in colour.

In people who were hypnotised, the part of the brain that processes colour vision changed when they received Spiegel's instruction. It became less active when he told them that a colour grid they were looking at was black and white, and more active when he told them a black-and-white grid was in colour. It was a crucial result, showing that the subjects didn't just pretend that the colour grid had been drained of its hues (or vice versa), they really saw it that way. This didn't happen in low-hypnotisable people, or in volunteers instructed to fake their response.

Hypnotised people behave differently too. Hypnotised subjects who have been asked not to see a chair that is in front of them will insist that it has disappeared. If asked to walk across the room, however, they will still move to avoid it (this fits the idea that their unconscious mind still knows the chair is there). By contrast, non-hypnotised subjects asked to fake the experience generally walk into the chair.

Thanks to research like this, doctors do generally acknowledge that hypnosis can reach beyond our conscious awareness to deep-seated thought patterns and beliefs. Hypnosis is recognised as a legitimate medical tool by the British and American Medical Associations, at least as a treatment for psychological problems such as addictions, phobias and eating disorders. But I'm interested in whether hypnotic suggestions can directly affect the physical body – particularly in a way that's medically useful.

Remember paediatrician Karen Olness, who treated Marette's lupus with cod liver oil and rose perfume? She is now a respected hypnosis researcher, and among other positions has served on the NIH Council for Complementary and Alternative Medicine. She claims that hypnosis helps us to reach the same unconscious parts of the brain as conditioned responses do, tapping into the autonomic nervous system to influence physical systems that aren't usually under voluntary control.

Her research with children shows that they can voluntarily influence blood flow to change the temperature of their fingertips.[6] Although fingertip temperature tends to increase when we're relaxed, 'these children were capable of increasing peripheral temperature way beyond what would be achieved merely from relaxation,' she says.[7] 'They would create different images. One of them said he was imagining that he was touching the sun.' Olness believes that mental images, so vivid when we are hypnotised, are crucial for influencing the physical body. Perhaps such images activate different parts of the brain than those associated with abstract or rational thought. 'But we're a long way from specifics on that,' she admits.

The finding that hypnotic suggestions can influence body temperature and blood flow has been replicated by other researchers, including Edoardo Casiglia, a cardiologist from the University of Padua in Italy. In one test, he told hypnotised volunteers that he was taking half a pint of blood from their arm. They responded with lowered blood pressure and constricted blood vessels, just as occurred in a second group who actually did give blood.[8] In another experiment, he told volunteers that they were sitting in a warm bath. Blood vessels across their whole body dilated as if they were sitting in a real bath; when volunteers were told that their forearm was in warm water, blood vessels dilated just in that forearm.[9]

In a third study, Casiglia asked volunteers to place their right hand into a bucket of ice-cold water.[10] This is an extremely painful task that usually evokes a strong fight-or-flight response, including constricted blood vessels, raised blood pressure and a pounding heart. It is an instinctive reaction; the conventional medical view is that we cannot suppress it voluntarily. Yet hypnotised subjects told that their right arm was insensitive to pain completed the task without any physiological effects.

According to Casiglia, if such effects were better understood they could have a range of potential medical applications. We might use hypnosis to boost blood flow to the brain (protecting against cognitive impairment as we age); to the extremities (to help people with poor circulation in their hands and feet); or even to direct a toxic drug to a particular part of the body. At the moment, this last one 'is science fiction', Casiglia admits, but not completely inconceivable – he says he has recently found that hypnotised volunteers can increase blood supply to their intestines on demand.[11]

Lab studies from other teams have reported that relaxation suggestions made during hypnosis can influence a variety of immune responses associated with stress, reducing inflammation, for example, in medical students facing exams.[12] Meanwhile some small trials have hinted that hypnotherapy may improve auto-immune disorders such as eczema and psoriasis, that it can reduce the duration of upper respiratory infections, and even clear warts.[13] The results are mixed, however. Different studies tend to measure different aspects of the immune system, and no consistent picture has emerged. As with hypnosis research as a whole, meta-analyses generally conclude that there is too little high-quality research to draw any strong conclusions about its benefits, or about which techniques work best. To an outsider like me, trawling through the data is a frustrating experience; despite glimpses of exciting potential, it's a field that mostly feels wishy-washy and obscure.

And then there is hypnotherapy for IBS.

***

Whereas many hypnotherapists delve into people's childhoods or psychological hang-ups, Whorwell wasn't interested in fixing his patients' personal problems. He wanted to target what he saw as the root cause of their misery: the gut.

The brain and the gut are intricately connected, he tells me. There's constant two-way communication between them, via the hard-wired connections of the autonomic nervous system as well as hormones that circulate through the bloodstream. Signals regarding what's happening in the gut travel to the brain, which then modulates gut function in response to that information – usually without us being aware of it.

For example, signals from the stomach tell us if we're hungry and need to eat; if we're full and need to secrete stomach acid or divert blood flow to aid digestion; or if we have ingested a poison and need to vomit. At the other end of the process, signals from the colon and rectum tell us when we need to poo. We can then either give the go-ahead, or suppress the impulse until a more convenient time.

Most of us have experienced how our state of mind can affect gut function. If we're uncomfortable with toilet arrangements we might not go for days, whereas when we're nervous we get butterflies or empty our bowels. 'The evolutionary value is that if you are roaming around in the savannah and something is about to eat you, it is good to empty your gut quickly so blood flow to your gut reduces,' says Whorwell. 'Then you can put all your blood flow to your muscles so you can run.'

In IBS patients, however, the communication between brain and gut goes haywire. Chronic stress, for example, can lead to persistent diarrhoea, vomiting, or painful gut contractions. This can create a vicious cycle in which people worry about their symptoms, making the problem even worse. 'The pain comes, then the anxiety comes,' says Emma, the 21-year-old who visited Whorwell's clinic with her mother. 'I know how it works, but I just can't break the cycle.'

After his training in hypnotherapy, Whorwell believed that the technique might reduce that stress and anxiety, helping patients not to over-react to signals from the gut. But he also hoped to influence gut function directly. To do this, he gave patients a tutorial on how the gut works, then during hypnosis asked them to visualise a calmer, trouble-free digestion process over which they had control. A popular method was to imagine the gut as a river. Someone with constipation might conjure a surging waterfall, whereas a diarrhoea patient might prefer boats on a slow-moving canal.

To survive his foray into hypnotherapy with his reputation intact, Whorwell knew he would have to document his results in robust scientific trials. He published the first one in 1984. It was a randomised trial of 30 people, who received 12 weekly sessions of either gut-focused hypnotherapy or psychotherapy (which involved discussing stress and emotional problems that might be contributing to their symptoms).[14] These were desperate patients who had suffered from severe IBS for years, with no relief from conventional treatment. He asked them to score their bowel function on a 21-point scale, with higher scores denoting worse symptoms. The psychotherapy group started with an average score of 13 and were no better three months later. The hypnotherapy group started the trial on 17, and finished it on one.

That's when a tentative experiment became a life's calling. Determined to drag hypnosis kicking and screaming into scientific acceptance, Whorwell has since set up a dedicated hypnotherapy unit at Wythenshawe Hospital, which now has six therapists, and has built up an impressive body of evidence supporting his technique.

Gut-focused hypnotherapy doesn't help everyone. Emma has been through the course, for example, and still suffers terribly. But over multiple trials and audits, Whorwell has shown that hypnotherapy helps 70–80% of patients for whom all other treatments have failed.[15] Other symptoms such as headaches and fatigue are eased as well as gut-related ones, and after hypnotherapy patients make fewer visits to doctors and consultants – not just for their IBS, but for everything. Small trials suggest that the approach is helpful for other functional gastrointestinal disorders too, including functional dyspepsia and non-cardiac chest pain,[16] and it may even help patients with more serious autoimmune disorders such as Crohn's disease and ulcerative colitis, in which the immune system attacks the gut lining.[17]

For IBS at least, the benefits seem to last long-term – when Whorwell followed more than 200 IBS patients who had responded to hypnotherapy for up to five years, 81% of them stayed well, in fact most of them continued to improve.[18] This lasting effect, and the fact that in trials patients receiving hypnotherapy improve significantly more than those in control groups, suggests that it isn't simply working as a placebo.

Although IBS patients can experience dramatic placebo effects, as we saw in Chapter 2, these are often temporary. Whorwell notes for example that when his patients have surgery they often feel better at first, but then relapse. By contrast, he believes that hypnotherapy helps to change patterns of thinking about their gut in order to ease symptoms for good. He gives patients CDs of their sessions, so they can keep practising at home as long as they need to.

Whorwell's studies also help to show that the therapy does more than reduce stress. In IBS patients, the gut lining is over-sensitive to pain, something you can measure by putting a balloon up someone's bottom and inflating it until they say it hurts. Healthy people feel pain at a pressure of around 40 mm Hg; IBS patients don't usually get to half that. Hypnotherapy seems to correct that hypersensitivity. When Whorwell tested them after a course of treatment, they were back into the normal range.[19]

And crucially, when patients are hypnotised, they are able to influence the speed at which the stomach empties its contents into the small intestine (measured using real-time ultrasound imaging),[20] as well as the rate at which the colon contracts.[21] As with Olness and Casiglia's experiments on blood flow, these are not things we are supposed to be able to do at will.

'You can't just sit there and tell the patient, "You've got to relax your muscles,"' says Whorwell. 'But in this hypersuggestible state, people seem to be able to do things to their body which they can't necessarily do in the conscious state.'

***

In Cruickshanks' card-lined office, I ask former flight attendant Nicole how she felt while she was hypnotised. As if she's floating, she says. 'When Pam's talking, I'm visualising warm, green-turquoise water. Soothing, sunshine holiday water. I feel like I'm smiling inside.'

And is the hypnotherapy helping her? She struggled to get it at first, she says. But since last week . . . She pauses, looking at us both, eyes bright like someone with a thrilling secret to share.

'A miracle has happened,' she says. 'The bloating and swelling was right up to my breasts. The pain was constant. Now, I've no bloating. I'm not taking any pain relief.' She turns to Pam, on the verge of tears. 'I want to kiss you! I've suffered so long. For me to say I've had no pain in one week – it's wonderful.'

Before Nicole leaves, Cruickshanks asks what her week has been like. 'I've just got cancer again,' she says, calmly. It's a tumour on her back. She's had it before, but now the cancer has returned. 'I'm so sorry,' I say, but Nicole shakes her head. 'It was caught early,' she says. 'I'm fine about it.' Then she points to her stomach. 'This is the worst thing. This is the most painful, soul-destroying thing.'

As she stands to go, she gives Cruickshanks a huge hug. Soon there will be one more thank you card on the wall.

***

Back in Peter Whorwell's office after my visit to the hypnotherapy unit, he wants to emphasise that IBS is not all about stress and anxiety. Other factors include genes, diet, gut microbes, the way the brain processes pain and, of course, the gut itself.

Every patient represents a different combination of these factors, he points out. In some cases, like Emma's pain or Nicole's bloating, psychology seems to play a large role. In others, like Gina's constipation, it may not be a significant factor at all.

He thinks Gina's problems have more to do with repeated abdominal surgery, which can damage the nerves necessary for the gut to function. As well as her hysterectomy, and having her gall bladder removed, 'She's had several rounds of surgery on her bottom,' he says. 'No wonder it's not working properly.'

This is why he insists that hypnotherapy should always be used alongside conventional treatment approaches. Although hypnotherapy might help Gina to manage the stress associated with her symptoms, Whorwell has also recommended powerful muscle relaxants and laxatives, and if they don't work, a colostomy.

I'm struck by how many patients referred to Whorwell have previously had abdominal surgery – at least seven of the ten patients I've met that day. It's a big factor in IBS, he confirms. If the gut is moved or disturbed during surgery it can become sensitised, and starts sending amplified pain signals to the brain. This is often what triggers IBS in the first place. In other cases gastroenterologists operate in the hope of alleviating patients' symptoms, only to find that their condition ultimately gets worse.

'Surgeons are programmed to operate,' says Whorwell. 'And in a lot of instances they bring about miraculous cures. If you have appendicitis or cholecystitis or a perforated bowel, they'll save your life.' But when somebody has abdominal pain, their default reaction is to remove something. Unfortunately, this often exacerbates the problem. 'It is done for the best reasons in the world,' says Whorwell. 'But once you have structurally changed the gut, causing scarring and adhesions, you are not going to be able to hypnotise that away.'

It reminds me of the dilemma faced by patients with chronic fatigue syndrome, who are pushed between CFS being either a biological, incurable disease, or a psychological invention. Are IBS patients too, I ask Whorwell, being caught between the two extremes of body and mind? Some are treated as if their IBS is a purely physical problem, with surgeons cutting out piece after piece of their bowel, while others are told the problem is all in their head. When what they really need is an approach that treats mind and body together?

Whorwell looks at me for a moment. 'Absolutely spot on,' he says.

***

You'd think that with all that he has achieved, Whorwell might be feeling pretty pleased about his career choice. He has developed a highly effective therapy, and helped thousands of patients whom other doctors had given up on. Teams around the world are carrying out randomised controlled trials into gut-focused hypnotherapy, also with positive – if not always quite so dramatic – results,[22] and a recent systematic review concluded that the treatment is effective and long-lasting.[23]

Thanks to evidence like this, the UK's National Institute for Health and Clinical Excellence (NICE), which approves medical treatments for use by the National Health Service (NHS), now recommends hypnotherapy for IBS where conventional treatments have failed. This is one of the only complementary therapies backed by NICE, and its only recommendation of hypnotherapy for a physical condition.

But Whorwell doesn't seem happy. Actually he seems quite disappointed. Because despite those trials, and that NICE recommendation, many of the administrative bodies responsible for funding treatment in the UK still refuse to support it, while the NHS website advises patients that research studies into hypnotherapy for IBS 'do not provide any strong evidence for its effectiveness'.[24]

According to Whorwell, one problem is that hypnotherapy isn't amenable to the strict trial designs that were developed for testing drugs. Before recommending a particular therapy, advocates of evidence-based medicine look for double-blind trials, where neither the patient nor their doctor knows whether they are receiving the real treatment or the fake one. This makes sense when testing drugs, in order to rule out that they aren't simply triggering a placebo effect.

But you can't hypnotise someone – or be hypnotised – without knowing. So reviewers or funders may look at the data on hypnotherapy for IBS, see that there are no double-blind trials and conclude that the evidence for it is poor. 'It's nonsense,' says Whorwell. And while it makes sense to blind patients in a drug trial, in order to separate the chemical action of the drug from any psychological effects, it misses the point when testing therapies such as hypnosis, when patients' beliefs and expectations are integral to how they work.

Whorwell argues that reviewers should be willing to accept evidence from a broader range of trial designs that are appropriate for testing a mind–body therapy but are still as close as possible to the gold standard. For example, researchers can carry out a single-blind trial, in which hypnotherapy is tested against a suitable control group and patients' symptoms are independently assessed by a researcher who doesn't know which treatment patients have received.

Jeremy Howick, an epidemiologist and philosopher of science at the Centre for Evidence-Based Medicine in Oxford, agrees that carrying out double-blind trials can be difficult or impossible for mind–body therapies, but points out that this is a problem for some conventional therapies too, such as surgery or physiotherapy. He suggests that in such cases it makes sense to forget the placebo group altogether and instead compare a therapy against other treatments that are known to be effective. 'If you have a health problem, what you want to know is what's the best treatment from all these alternatives?' he says. 'That's what patients care about.'[25]

A deeper problem may be that hypnotherapy is very unfashionable in most scientific and medical circles, and still carries those connotations of quackery. Proponents complain that there is very little funding available for research into hypnosis, even compared to other mind–body therapies such as meditation,[26] and little interest in studying how it might help patients. 'The majority of health care professionals just don't think that it's necessary or important,' says hypnosis researcher Karen Olness.

Over the years, Whorwell has attempted to expand his hypnotherapy model beyond gastrointestinal disorders. He says he has approached specialists in a variety of fields, thinking that the technique might help patients to manage the pain and anxiety associated with conditions from eczema to cancer. All turned him down, including one who told him, 'I don't think what you are doing could possibly help any of our patients.'

'There is tremendous prejudice against hypnosis,' Whorwell concludes. 'Medicine has become terribly technical. We're wedded to drugs, scans, all this high-tech stuff. Something as simple and mundane as hypnosis can't be seen as being any good.' Embracing hypnotherapy would require rethinking not just trial design, he says, but how to do medicine. 'The standard medical model of treatment is take a history, give them a drug, send them away, if the drug doesn't work, give them another drug and so on. This is a different model where you throw away the prescription pad, you throw away the desk, you throw away everything and you are the thing that either makes them better or not.'

Whorwell has just published an audit of another thousand patients treated with gut-focused hypnotherapy.[27] He reels off the stats: 76% with a clinically significant reduction in symptoms; 83% of responders still well after one-to-five years; 59% taking no medication, 41% taking less; 79% consulting their doctor less often or not at all. He is due to retire soon, though, and is planning no more trials. 'I think we've probably blown it by then,' he says.

'We have produced a lot of good research, incontrovertible research. Yet we're always fighting the people who fund treatment. They're always saying there's not enough evidence. How much more evidence do they want?'

Perhaps he's right, and the barriers to acceptance for the therapy, with its chequered history, are just too strong. But across the Atlantic, hypnosis is being reinvented yet again.

6

RETHINKING PAIN

Into the Ice Canyon

I'm floating slowly through a shimmering ice canyon. The walls are sheer and there's a ribbon of blue water below. Perched on icy shelves either side of me are penguins, waving their flippers, and snowmen with smiles made of coal. I throw snowballs at them, and if I get a direct hit they explode in a flurry of triangular shards, leaving their smiling faces in the air like a row of frozen Cheshire cats. In the background, Paul Simon is singing 'You Can Call Me Al'.

I look up towards snowflakes and a dark sky; down to the water; then spin around. But mostly I let myself drift forwards. There are ice bridges to watch for, some supporting glossy igloos. The snowmen start throwing snowballs back at me, so after a while I stop trying to hit the men and aim for their missiles instead, setting up satisfying mid-air collisions.

As I fly around a bend I see a family of woolly mammoths with huge curved tusks, standing knee-deep in the water below. I fire a snowball at one and he trumpets. Then some flying fish appear, silver blue, leaving trails of snowflakes as they leap downstream.

At various moments as I progress down the canyon, I become vaguely aware that something is happening to my foot. There's tingling, then something that if I thought about it might be burning pain. But that's in another world, of no relevance to this magical canyon, and I can't be bothered to focus on it right now. I'm more interested in whether I can get those mammoths to explode.

***

In 2008, Lieutenant Sam Brown was deployed for his first tour of duty, to Kandahar in Afghanistan. At dusk on the last day of his mission, a call came through from a nearby platoon to say that they had been ambushed. Brown led his men through the desert to help, but on the way his Humvee hit a roadside bomb.[1]

He saw a bright flash as the armoured vehicle lifted into the air; seconds later, it was a twisted pile of wreckage. He doesn't remember how he got out, but his body was on fire. He thought he was going to burn to death by the blast crater, but with the help of his gunner, he smothered the flames with handfuls of sand. By the time they had put out the fire, the sleeves of his uniform had burned off and the skin was gone from his body, face and hands. The flesh left behind was raw red or charred black.

Brown was eventually airlifted to Brooke Army Medical Center in San Antonio, Texas. He had suffered third degree burns over much of his body. Also referred to as full-thickness burns, third-degree burns destroy all layers of the skin. Doctors kept him sedated for weeks while they harvested skin from his back and shoulders to cover the worst. He woke to a series of further surgeries, including the amputation of his left index finger. But the hardest thing was enduring the daily sessions when nurses scrubbed dead tissue from his raw wounds. It was like being burned all over again.

Later, as his burns started to heal, he needed daily physiotherapy, which turned out to be even more painful. Over wounds as extensive as Brown's, scar tissue tends to thicken and contract. To ensure that he would still be able to move freely once his burns had healed, his therapists had to force his body and limbs past their limits, to stretch and tear the scar tissue as it formed.

In the United States each year, an estimated 700,000 people visit emergency rooms for the treatment of burns, of whom 45,000 or so are admitted to hospital.[2] To help them through their gruelling wound care and physiotherapy sessions, they are given opiate drugs at some of the highest doses used in medicine. But the amount that doctors can give is limited by their side effects – from itching and being unable to pee to loss of consciousness and death. At the highest safe doses, many patients are still left in agonising pain. And taking opiates for months puts them at risk of addiction.

Brown tried desperately to limit the drug doses he was taking. He found the physiotherapy so unbearable that at times, his superior officers had to order him to undergo treatment. But he feared becoming an addict even more. Then he was asked if he wanted to take part in a pioneering research trial.

***

There's no shortage of painkilling drugs in healthcare. We have over-the-counter tablets such as aspirin and ibuprofen; potent narcotics such as morphine and codeine; sedatives such as ketamine. Antidepressants, anticonvulsants and corticosteroids can all be used for relieving pain. Doctors can anaesthetise a small spot of skin, an entire region of the body, or render a patient completely unconscious. Unfortunately, none of this means we have eliminated pain in medicine. Not even close.

Pain is a particular problem for people undergoing medical interventions and procedures during which they have to be awake – wound care for burns patients like Brown, for example, or keyhole surgery, which is increasingly replacing open surgery for everything from biopsies and diagnostic tests to inserting medical devices and destroying tumours. As Brown's case shows, pain medications on their own often don't do enough, leaving even drugged-up patients in agony.

Then there are the millions of people affected by chronic pain, in conditions from arthritis to fibromyalgia. Over the last couple of decades there has been a surge in the amount of opioid drugs such as oxycontin – artificial equivalents of the endorphins involved in the placebo effect – prescribed for such conditions. These used to be drugs of last resort, used only in severe cases such as terminal cancer. But they are now routinely prescribed in patients with mild to moderate pain, who can end up taking them for months or years.

The trouble is, unlike natural endorphins in the brain, these artificial versions swamp the brain's opioid receptors. In response, the receptors become less sensitive to the drug. We develop tolerance, needing higher and higher doses for the same effect. It also means the drugs are terribly addictive. Coming off them leaves people with horrible withdrawal symptoms including anxiety and hypersensitivity to pain, because their desensitised receptors no longer respond to natural endorphins as they should.

The increase in prescriptions has led to a surge in opioid-related addictions and fatal overdoses that has been described as 'one of the great unfolding tragedies of our time'.[3] This is a particular problem in the US, which makes up less than 5% of the global population but consumes 80% of the world's supply of opioid prescription drugs.[4] By 2012, 15,000 Americans were dying each year from prescription pill overdoses, more than from heroin and cocaine combined.[5] In 2013, the US Centers for Disease Control and Prevention (CDC) named painkiller addictions the worst drug epidemic in US history.[6]

Which begs the question, are we approaching pain all wrong? Instead of prescribing higher and higher levels of addictive painkillers, some researchers claim there is another way. They are harnessing the power of illusion to reduce drug use and ease our pain.

***

When I arrive at the experimental pain lab at the University of Washington Medical Center in Seattle, I'm greeted by research assistant Christine Hoffer. She asks me to remove my right shoe and sock, then straps a small black box tight against the skin of my foot. It's designed to inflict pain, she explains, by rapidly heating up. Usually Hoffer gives her volunteers repeated electric shocks too – but luckily for me that equipment isn't working today.

She activates the box for 30 seconds, and asks me to score how painful it is on a scale of one to ten. Then she increases the heat in half-degree increments, looking for a response roughly in the middle of the scale. Eventually I score 6 out of 10 for both intensity of pain and unpleasantness. It's a stinging, burning sensation, not enough to leave a blister but intense and impossible to ignore. This is the temperature that Hoffer will use for the experiment.

She fits me with virtual reality goggles that project high-resolution 3D images, and noise-cancelling headphones with surround sound. Suddenly I'm floating in the snow, admiring the sparkling walls of a canyon made of ice. Hoffer shows me how to move around and fire snowballs with a computer mouse. The graphics are cute but not super-realistic, particularly by the standard of many of today's video games. Yet there's an immersive quality that I've never experienced before. All sights and sounds from the outside world are blocked out, and as I look around, the virtual world extends above, below and behind me. Cartoonish as the landscape is, I feel that I'm inside it.

I spend ten minutes with the snowmen and penguins, during which Hoffer turns on the heat box three times. Afterwards she asks me to score the experience again. My pain intensity score comes down a bit, from six to five (but each time felt as a brief peak rather than the longer plateau I endured before). Meanwhile the unpleasantness of the pain falls dramatically from six to two. I score my overall enjoyment of the experience as 8 out of 10, pretty fun, and would have been happy for another go.

It's all about attention, says anaesthesiologist Sam Sharar, who runs this lab. The brain has a fixed capacity for conscious attention. We can't increase or decrease it, he tells me, but we can choose what we pay attention to. If we focus on a painful sensation, it will increase our experience of that pain. But if we think about something else – something safe, pleasant, far away – the pain we feel is dimmed.

Visual imagery is a particularly potent form of distraction. Sharar shows me a video of the hiker Aron Ralston – who was forced to amputate his own forearm with a pocketknife after five days trapped in a remote Utah canyon in 2003 – describing afterwards how mental images helped him to survive his ordeal.[7]

On his fifth night in the canyon, Ralston was shivering with cold, badly dehydrated and in agonising pain from his hand, which was crushed beneath a fallen boulder. He knew that he was going to die. Then he saw a vision that blocked out his traumatic surroundings. 'There was a little boy about three years old,' recounts Ralston. 'He was wearing a red shirt, and he was playing with a truck, moving it around, making little brrm brrm noises.

'Then he stopped, and he looked over his shoulder, and he came running over to me, and I could see myself pick up this little boy and then boost him up there on my shoulder, where we were looking straight into each other's eyes. And I knew that I was seeing the face of my future son. Then that vision blacked out and I was back in the canyon, shuddering from that hypothermia.'

Ralston goes on to say that imagining his loved ones helped him to tolerate the pain of cutting off his arm. 'As I cut it, I felt the worst pain that I have ever experienced in my life. For 30 seconds, all I could do was close my eyes and breathe. But I never said ow, I never shed a tear, I never cried. And that's not because I'm superhuman. It's because when I opened my eyes, all I could think about, all I could imagine, was seeing my family again.'

For Ralston, an internally generated world – images of his family and imagined future son – helped to focus his attention away from the pain of his horrific ordeal. The virtual ice canyon I've just experienced, says Sharar, is an attempt to create artificially that same effect.

It's the brainchild of Hunter Hoffman, a cognitive psychologist at the University of Washington who specialises in building virtual worlds. Back in the 1980s, Hoffman had just created 'kitchen world', a virtual kitchen fitted with countertops and cabinets as well as objects you could pick up, like a teapot, toaster, frying pan – and a wiggly-legged spider in the sink. Hoffman hoped to help people with arachnophobia, by giving them a safe place to get used to contact with spiders.

Then he heard from a friend about the work of David Patterson, a psychologist who was using hypnosis to ease the pain of burns patients at UW Harborview Medical Center, also in Seattle. The friend thought that the technique might have something to do with distraction. I've got a distraction for him, said Hoffman, and the pair started working together, to see if virtual reality (VR) could help patients going through some of the most painful procedures in medicine. First they put them inside kitchen world. 'It worked with the very first kid,' says Hoffman. So he set about designing a virtual world just for burns patients.[8]

At the time, making any kind of virtual world was at the cutting edge of technology. Hoffman used a supercomputer made by the company Silicon Graphics, which cost $90,000 including a heavy helmet, and based his new landscape on military flight simulator software that modelled a fighter jet taking off from an aircraft carrier. It needed a few adjustments. 'We were super worried about simulator sickness,' he tells me. 'A lot of burn patients are nauseous from their pain meds. I was convinced from the very first patient that VR had the potential for pain distraction, but I was worried that nausea would be a showstopper.' So he closed in the terrain to a narrow canyon to discourage people from changing direction or spinning in circles. And he built it out of soothing ice. He called it Snow World.

Twenty years later, the essence of Snow World is still the same. But the supercomputer and helmet have been replaced by a laptop and hi-res goggles (helmets are no good for people with burns on their heads and faces). Hoffman has designed electricity-free fibre-optic goggles, which carry signals down 1.6 million tiny glass fibres per eye, so they can be used in water tanks while patients have their burns scrubbed. He has also upgraded the graphics, and changed the background music. Paul Simon once tried Snow World at an exhibition, Hoffman explains. He loved it but hated the ethereal, spacey music they were using, so he donated his own.

The UW team has also carried out a series of randomised controlled trials on healthy volunteers (with Hoffer's heat box and electric shocks) and on burns patients at Harborview. They've found that Snow World works strikingly better than other forms of distraction such as music alone, or video games. The essential ingredient seems to be how immersed you feel in the world. The greater the sense of presence, the more pain relief people feel.

In the lab, Snow World consistently cuts pain scores by 35%, says Hoffman, compared to around 5% for music. And when used in combination with pain medication, it reduces patients' pain ratings by an extra 15–40% on top of what they get with drugs.[9] The researchers see the effects not just in subjective pain scores but in brain scans too, with activity in pain-related brain areas almost completely extinguished.[10]

The team is still experimenting with ways to boost the effect – small doses of the hallucinogenic drug ketamine seem to improve the sense of immersion people feel, for example. But the Snow

World technology is already being used in around 15 hospitals across the US. One of them is the Brooke Army Medical Center (BAMC), in Fort Sam Houston, Texas, which has treated hundreds of soldiers burned in combat in Iraq and Afghanistan. Most of them suffered burns from improvised explosive devices (IEDs) – roadside bombs, car bombs, suicide bombs, or as Hoffman puts it: 'These real fancy bombs that would really blow up a Humvee super-bad.'

Hoffman and his colleagues carried out a trial of 12 soldiers at BAMC, including Lieutenant Brown.[11] When they were immersed in Snow World during their physiotherapy sessions, their worst pain score dropped nearly two points compared to the part of the session spent without it. The proportion of time they spent thinking about the pain dropped from 76% to 22%. And whereas they rated their normal physiotherapy as 'no fun at all', they rated therapy while in Snow World as 'pretty fun'.

Snow World worked best for the six patients who started off with the worst pain; the soldiers who needed it most. For example Brown's worst pain score dropped from 10 to 6, and when in Snow World he rated his therapy – previously so gruelling – as actually quite fun. He later told a reporter for *GQ* magazine that it reminded him of skiing with his brother during Christmas break in Colorado, back when he was still a cadet at the US military academy at West Point.

After the session he gave Hoffman his verdict: 'I think you guys are onto something.'[12]

***

One night in April 2014, 22-year-old Terrell was driving at 80 mph down a high street between Kent and Des Moines, just south of Seattle, when he lost control of the car. The vehicle flipped, did two turns in the air and skidded to a halt. Then it caught fire.

An ambulance brought Terrell to Harborview Medical Center

with a broken arm and serious burns to his leg and chest. 'When I woke up, I was in the worst pain,' he tells me. 'There were tubes in my throat, tubes everywhere. I was trying to get them out, they were stopping me. My face was swollen.' There were burn marks all over his body. Once Terrell had calmed down he called his girlfriend to tell her he had been in an accident. 'She didn't believe me,' he says. 'But when she got here, then she knew.'

A month after the crash, Terrell is lying in his hospital bed, dressed in a green robe with gathered frills at the shoulders and propped up on about five pale blue pillows. He is slightly built, with a tuft of a beard on his chin and unshaven sideburns. Two coin-sized scars gleam white against his dark skin, next to his right eye and on his forehead. His left leg is heavily bandaged, with yellowy brown serum seeping through at the foot.

He's surrounded by the remains of unfinished meals – milk cartons, a nibbled muffin, a dinner plate, yoghurt pots and empty cups – and a bunch of helium balloons with shiny foil messages: 'You're so special' and 'Get well'. A few feet away on the other side of a curtain is a huge, angry-looking man; his scowling face is burned pink and brown and his bandaged arms stick straight out to either side. He seems to have enemies outside the hospital; his name has been removed from his hospital records for his own protection, one of the medical assistants whispers to me as we pass.

Over the past few weeks, Terrell has had four or five surgeries (he can't remember which), to graft skin from his right leg onto the burns of his left. He's still on hefty doses of the opioid drugs methadone and hydromorphone for his pain, which make him permanently drowsy. When the anonymous man starts shouting, 'My pain is at ten, someone get down here now!' I struggle to hear Terrell's soft, slurred speech.

He tells me he's from Renton, a city just south of Seattle, where he lives with his mother and his girlfriend. I ask what Renton is like and he says there are 'some dangerous people' and that he didn't finished high school because he was 'being bad'. He's currently unemployed, but when he gets out of hospital he hopes he might get a job at the fast-food chain Popeye's, washing dishes: 'They hire felons and people like that.'

Tattoos cover Terrell's arms and chest. Among the swirling faded designs, I make out an empty-eyed clown face and several figures with bared teeth and protruding ribs. He dismisses them – 'just art', he says. Small letters on his right arm read 'Son of God', while larger initials on his left spell out 'M.O.E.' His girlfriend? No, he laughs. 'Money over everything.'

An assistant wheels in a clunky grey cabinet carrying a laptop and a set of goggles. Terrell settles back on his pillows with the headset and phones, while the open laptop reveals what he's watching.

It's just like the equipment that transported me to Snow World, but this is a very different scene. Terrell is floating along a stream, a rocky trickle at first, which gradually opens out into clear, shallow river with sandy banks. On either side there's grass then a dense forest of pine trees. Straight ahead, snowy-topped mountains are visible beneath a clear blue sky. This isn't a game; there are no penguins or snowballs to shoot. Instead, it's a session of hypnosis. The numbers one to ten float past, then a soothing male voice delivers suggestions for feeling relaxed and free of pain.

Terrell has never heard of hypnosis. But two days ago, after he complained to staff that his pain was 'a ten' despite the drugs he was taking, they asked if he wanted to try a relaxation aid and he said yes. 'When I did that I couldn't feel no pain,' he says. 'I wasn't worrying about it.' Today, he's keen to try it again. Terrell lies still as the programme runs, at first absorbed in the peaceful forest scene. But then his eyes close, and his mouth drops open. He's asleep.

***

It's a common problem, says Hoffman's psychologist colleague David Patterson, when I recount this story to him later. Patterson has worked with burns and trauma patients at Harborview for the past 30 years, looking for non-pharmacological methods to ease their pain beyond the relief they get from drugs. Although Snow World is extremely good at distracting patients from their pain for short periods of time, the effects disappear as soon as they take off the goggles. So Patterson is also investigating whether positive suggestions delivered by hypnosis can reduce pain and aid their recovery in the longer term.

The idea of using hypnosis as an anaesthetic was pioneered by James Esdaile, a Scottish surgeon working in India in the mid-nineteenth century. He saw thousands of patients affected by lymphatic filariasis, a parasitic infection that causes huge fluid-filled swellings, but had difficulty persuading sufferers to let him remove these protuberances. At the time, there were no available anaesthetic drugs. Without them, the operation was excruciatingly painful, and many patients died from the shock.

Esdaile had read about the analgesic effects of mesmerism, which was popular back in Europe at the time. Although he had never seen anyone being mesmerised, he decided to give it a try and was surprisingly successful. The surgeon kept detailed notes of the patients he operated on, including a 40-year-old shopkeeper called Gooroochuan Shah, who had a giant, 80-pound scrotum that he used as a writing desk.

Esdaile cut off the monster swelling after Shah was rendered 'insensible' by mesmerising him, and was convinced that the procedure saved the man's life. 'I think it extremely likely,' he wrote, 'that if the circulation had been hurried by pain and struggling, or if shock to the system had been increased by bodily and mental anguish, the man would have bled to death.'[13] As the word spread, patients with lymphatic filariasis flocked to see Esdaile, and his hospital became a kind of 'mesmeric factory' in which he carried out thousands of operations with very low death rates for the time.

Today, Esdaile's techniques have been largely forgotten. Now that we have effective chemical anaesthetics, most of us have no need to undergo surgery drug-free. (There are many situations, however, in developing countries and in war and disaster zones, where this isn't the case. Four thousand people had limbs amputated after a devastating earthquake hit Haiti in 2010, for example, mostly without any form of pain relief.) But a few researchers are pursuing whether hypnosis can help to reduce drug use for wound care, recovery from surgery and chronic pain.

Patterson tells me that he became interested in hypnosis after a 'life-changing' experience within a few months of starting on the burn unit at Harborview.[14] A badly burned patient in his sixties was struggling to cope with his wound-care sessions. 'He was maxed out on every drug – morphine, tranquilisers. He said, "I can't go back in there, I'd rather die."' Patterson's mentor, a pain psychologist called Bill Fordyce, suggested that he try hypnosis.

So Patterson found a script for inducing hypnosis in a book, and read it out to the patient. It was designed so that when the nurses later touched the man on the shoulder during his wound care, he would go into a trance. 'When I went back to see what had happened, the ward was buzzing,' says Patterson. 'They said, "What did you do to that guy? We touched him on the shoulder and he fell asleep." It was astounding.'

Since then, brain-scanning studies have revealed that suggestions of pain relief delivered under hypnosis influence areas of the brain involved in pain perception. And several small, randomised controlled trials suggest that adding hypnosis to conventional treatment significantly reduces chronic and acute pain in a range of conditions.

The trouble is, most of the people whom Patterson sees are not straightforward to hypnotise. Harborview caters to all major traumas and burns cases in the region, from gunshot wounds to car accidents, regardless of whether they have medical insurance.

Many patients there have psychiatric problems, or are addicted to alcohol or drugs. And like Terrell, they're generally in pain and doped up on powerful painkillers, which means they're sleepy and find it hard to concentrate, and they may have no idea what hypnosis is. They often aren't able or willing to focus on a traditional hypnotic induction.

Another downside to conventional hypnosis is that it can be expensive, because you need a staff member to deliver it. So Patterson wondered if he could solve both problems by using virtual reality to immerse patients in a hypnotic trance. With a pre-recorded virtual session, patients don't have to generate their own visual imagery, and the treatment can be delivered anywhere, anytime, without a live hypnotist.

The first patient Patterson tried it on, in 2004, was a 37-year-old volunteer fireman called Grant. Six weeks earlier, Grant had poured gasoline into a barbecue pit, not realising that it still contained the embers of a previous fire, and in the resulting fireball suffered deep burns to 55% of his body. Since then he had endured six agonising operations to graft skin onto the burns and was still in excruciating pain. Unless he was kept heavily sedated, he became delirious and suffered violent panic attacks, particularly during the daily sessions when medical staff needed to clean and dress his wounds. 'He was at his wits' end,' says Patterson. 'All we had was Snow World.'

Rather than an interactive game, Patterson asked Grant to watch a pre-recorded sequence. Floating igloos showed the numbers one to ten as he floated down through the ice canyon. At the bottom, Patterson's voice suggested to the patient that he would feel relaxed and pain-free during subsequent sessions of wound care.

On the first day of the study, before any hypnosis, Grant scored his pain at a maximum of 100, despite being on sky-high doses of painkillers – 15 times higher than the typical dose used for burns patients at Harborview. The next morning he watched a session of virtual reality hypnosis. During his wound care later that day, Grant's pain score came down to 60, and on the third day, after a top-up session of audio hypnosis, he rated his pain at just 40. In the meantime, the drug dose he needed came down by a third. On the last day of the study, Grant again had no hypnosis. His pain score shot back up to 100; in fact he was so distressed by the pain that he was unable to complete the rest of Patterson's questionnaire.[15]

Since that case study with Grant, Patterson has developed the relaxing forest scene for delivering VR hypnosis, and he has reported positive results in several other burns patients, as well as trauma patients like Terrell. In a pilot trial of 21 patients in severe pain from broken bones and gunshot wounds, Patterson compared VR hypnosis against the Snow World game, or no treatment.[16] The patients had a virtual reality session in the morning, then were asked to rate their pain for the rest of the day. After Snow World or no treatment, the patients' pain rose over the course of the day, whereas in the hypnosis group it went down.

Patterson is now carrying out a bigger trial of 200 trauma patients to compare VR hypnosis with audio-tape hypnosis and standard care. But for now, 'It's brand new,' he says. 'The jury's out.'

***

Here's something you can try at home. Place your right hand on a table in front of you. Keep your left hand out of sight underneath the table or behind a screen, and place a fake hand (a stuffed rubber glove will do) on the table in its place. Now ask a friend to stroke both left hands – the visible fake hand and the hidden real one – at the same time. After a few seconds, you should experience a strange effect; it will feel as if the rubber hand is in fact your own hand.

This is a phenomenon known as the 'rubber hand illusion'.

Even though you know that the fake hand isn't part of your body, you feel as if it is. Once the illusion is established, it affects brain activity and behaviour. People respond more quickly to objects that they see on or near the fake hand (just as with their own hand), and instinctively flinch or try to remove the hand if someone approaches it with a needle or knife.

But it also has physical effects. Neuroscientist Lorimer Moseley at the University of South Australia in Adelaide has recently shown that during the rubber hand illusion, blood vessels in the unseen hand constrict, decreasing the flow of blood to this body part and causing its temperature to drop. Allergic responses in the unseen hand also appear to be boosted in a way that's consistent with immune rejection.[17] It's as if the lost hand is no longer treated as such an integral part of the body.

This supports the claims of hypnosis researchers, described in Chapter 5, that through using suggestions and illusions it is possible to influence blood flow and immune responses. Moseley concludes from his studies that we all have a 'mind map' of ourselves – a mental representation of our physical body – held in the brain.[18] This keeps us updated about the extent of our bodies and where we are in space, and may also play a crucial role in controlling and regulating our physiology (including things like immune responses and blood flow). Changes to the mind map, in this case achieved through a simple visual trick, are felt not just in the brain but in the body too.

This could have major implications for our health. Moseley speculates for example that the brain's unconscious perception of different parts of the body might play a role in some autoimmune diseases. A mismatch between the mind map and reality can also be a cause of chronic pain – if sensory information coming from a particular body part clashes with what the brain expects, for example, it triggers pain to warn us of potential danger.

Phantom limb pain, in which amputees feel pain from a limb that no longer exists, is one obvious example, but problems with perceived ownership might be involved in other chronic conditions, such as complex regional pain syndrome (CRPS). Patients with this condition suffer intense burning pain after injuries such as wrist fractures, long after the bone itself has healed. The affected hand gets cooler in CRPS, just as it does during the rubber hand illusion.

Even relatively minor injuries can trigger alterations to the mind map as the brain struggles to interpret sensory information it receives, says Candy McCabe, a professor of nursing and pain sciences at the University of the West of England. 'Quite quickly you can move into a system where everything is healed up on the periphery, but the central nervous system becomes over-sensitised to things that shouldn't normally cause pain.'[19]

For example, in osteoarthritis, a condition caused by mechanical damage and inflammation in the joints, there's no close correlation between the degree of structural damage and how much pain people feel. What's often driving the pain, McCabe argues, is not the problem joint itself but how the brain *perceives* that joint. Just as with the central governor theory of fatigue, pain researchers are repeatedly finding that although messages from the body are important for pain, these are always modulated by our perceptions (conscious and unconscious) of how much danger we are in.

Researchers including McCabe and Moseley are now investigating whether tricking the brain into seeing a healthy limb can reduce pain in phantom limb syndrome, CRPS, stroke patients and osteoarthritis.[20] In a variation of the rubber hand illusion, they are placing patients in front of a mirror or screen so that instead of their diseased limb they see the reflection or image of a healthy one. Whereas the virtual reality hypnosis and distraction developed at Harborview create an overall illusion that we are in a safe place, perhaps mirror therapy can perform a more focused trick, convincing the brain that an affected body part is safe and well.

Unfortunately, despite the public health disaster being wrought by prescription painkillers, there is relatively little research interest in non-pharmacological methods to help people deal with pain, and as we saw with hypnosis research in the last chapter, the studies so far are small. A recent review concluded that there isn't enough high-quality evidence to say for sure that mirror therapy works better than placebo.[21]

Stanford hypnosis researcher David Spiegel suggests that part of the reason for the lack of enthusiasm is economic. Pain relief is a billion-dollar market, and drug companies have no incentive to fund trials that would reduce patients' dependence on their products, he points out. And neither have medical insurers, because if medical costs come down, so do their profits. The trouble with hypnosis and other psychological therapies, he says, is that, 'there's no intervening industry that has the interest in pushing it'.[22]

That could be about to change, however. In March 2014, Facebook bought a little-known California startup called Oculus for $9 billion. The company specialises in VR gaming and has just developed a headset called Oculus Rift, similar in size and shape to a scuba mask. Whereas the VR equipment that Hoffman and Patterson use costs tens of thousands of dollars, Oculus sells its headsets for just $350 each. That promises to bring virtual reality within reach of ordinary consumers, who will be able to run wireless masks from their tablets or smartphones. Hoffman says he's already tried running Snow World on an Oculus Rift headset, with a burns patient undergoing physical therapy. 'It worked real well,' he says.

Developments like this mean that people will soon be able to use virtual reality pain relief – whether distraction games, hypnosis, or mirror-type illusions – at home. It also means that virtual worlds are about to get much more sophisticated, predicts Hoffman, as video game companies throw resources at developing software to go with the new headsets. As well as better games, he says, that could lead to better pain therapies too. It also makes me wonder whether we might soon see pain relief trials funded not by drug companies, but by the gaming industry.

In the future, Hoffman envisions entire libraries of off-the-shelf virtual worlds that those in pain can choose to match their interests. And the possibilities go beyond pain relief – he's still interested in using virtual worlds to treat psychological disorders, for example, and has designed World Trade Center world, terrorist bus bombing world and Iraq world, to allow patients with post-traumatic stress disorder to face their fears.

Maybe virtual reality will even become powerful enough to shift attitudes in the medical community. 'VR distraction is valuable for patients now,' says Hoffman. 'But I think it has enormous potential for precipitating a paradigm shift in how pain is treated. The results are so strong, it's encouraging the medical community to start exploring the use of non-pharmacologic analgesics in addition to pain meds. Who knows where that is going to lead?'

***

Two days after our first groggy meeting, I go back to see Terrell, and I'm surprised to find him alert and smiling. He's got a shoe on his bandaged foot – 'I call it my "do anything shoe",' he jokes. He has just taken a shower on his own for the first time since his accident and has even been to the gym. Whereas the doctors had previously said that he would be in hospital for another two weeks, they have now promised that he can go home by Monday, in three days' time.

Does he think the virtual reality helped? Since trying it, his injuries still hurt. 'But I've felt a little bit different,' he says. 'More at ease.' This impression is confirmed by one of the nurses, who tells me that Terrell underwent a 'personality change' after his first session of hypnosis, from being downright sullen to polite and friendly.

When I ask him what he liked about it, he decides it was the trees. 'There ain't no better place than a forest,' he says. 'If you was mad, you could go to a forest and get all of that out.'

All of what? I ask. 'All of the pain.'

7

TALK TO ME

Why Caring Matters

I remember bright lights and Tom Jones (the surgeon's choice); a tall, blue screen across my chest; and talking to my partner about the first thing that came into my head – ice cream, as it turned out – to distract myself from the odd, rummaging sensations coming from inside my abdomen. Then a baby girl, drenched in blood, was lifted high above the screen.

It was August 2009. Days earlier, heavily pregnant with my first child, I hadn't been too worried about the birth. I was fit and healthy and had attended all my pre-natal classes. My local hospital had a midwife-led centre with birthing balls and water pools. I was excited about feeling the first contractions, and planned to sail through the delivery with some relaxing massages and deep breaths.

It didn't happen that way. For several days of early labour I felt no recognisable contractions, just a searing pain in my pelvis that left me unable to eat or sleep. Things just didn't feel right, and by the time I got to the hospital – high blood pressure meant I was ineligible for the birthing centre, so I ended up on the obstetrics ward – I was exhausted and scared.

A midwife promptly broke my waters, wired me up to a foetal heart monitor, and administered artificial oxytocin to boost my contractions. That's when I realised that the searing sensation I'd felt before had been mere discomfort. Now the dial jumped to ten – my pelvis was surely breaking, something must be wrong. Overwhelmed by fear and pain, I started to panic.

The midwife seemed frustrated. In her view I was still in relatively early labour, and should have been coping better. I've climbed mountains, I wanted to protest. I've dived with sharks (well, reef sharks at least). I have a black belt in jiu jitsu! I'm not a wimp with no willpower or tolerance for pain. But it's hard to talk when your awareness is dissolving into screaming white noise. This was all perfectly normal, the midwife insisted between contractions. Her words made me feel alone. Either she had no idea what I was experiencing, or I was a complete and utter failure at childbirth.

I found out much later that my baby was in a difficult position, facing forwards instead of backwards, which meant that instead of fitting smoothly into the birth canal, her skull jutted awkwardly against it. Given time, babies in this position sometimes turn. But when the midwife broke my waters and administered oxytocin, the fluid cushioning the baby's progress was gone, and my contracting uterus forced her skull inexorably downwards, bone grating against bone.

I requested an epidural, and the disappearance of the pain was magical. As sometimes happens after an epidural, however, my contractions slowed. I spent the next 24 hours flat on my back surrounded by wires, drips and monitors. With the first midwife long gone, a series of others came and went. They checked graphs and upped doses, carried out internal exams to check on my progress, and poked a needle into my baby's scalp to check on hers. Eventually, a doctor informed me that she was stuck and I would need an emergency caesarean section.

I didn't hold the baby at first; I was nauseous and shivering violently after the surgery, and no one thought it was a good idea. Without that initial contact, my daughter subsequently struggled to breastfeed. She started her life crying and hungry in a Perspex-walled cot (she lost over 10% of her body weight in her first week), while I was scolded day and night for her condition by a further carousel of midwives and health visitors.

One of them made me spend hours expressing drops of precious colostrum into tiny syringes (tricky in the dark when the only lamp is on the wall behind your head), then the next would come on shift and chide me for having left my baby in her cot. Another repeatedly folded my breast into my daughter's mouth as if she were stuffing a chicken. I wondered how long it is possible for a person to go without sleep.

Four days and several panic attacks later, I was allowed home. I was overwhelmingly grateful to have a healthy child, but I was left wondering whether there might have been another way.

***

In other words, it was a typical birth. Thanks to modern medical care, childbirth is now extremely safe. In the UK, only around 0.7% of babies are stillborn or die shortly after birth.[1] The proportion of women who die is even lower. And we have ready access to pain relief. Yet despite all that, labour is often a distressing experience. In one survey, nearly half of women interviewed two days after the birth of their babies said that it was the worst pain imaginable, even though 91% of them had received pain-relieving drugs.[2]

And many women are left with mixed feelings about the birth of their babies. Around a third of women feel traumatised after giving birth, while 2–6% of women suffer from full-blown post-traumatic stress disorder (with women who have experienced instrumental deliveries or emergency c-sections at increased risk).[3]

Meanwhile more than half of births in developed countries such as the UK and US are 'assisted', which means they are either induced or involve the use of instruments or surgery.[4] Such outcomes can have long-term health implications for mother and child. Take c-sections, for example. Particularly when surgery is carried out as an emergency, potential complications range from bladder damage and infection to life-threatening haemorrhages and blood clots.

Women who deliver by c-section also risk complications in future pregnancies, including uterine rupture and problems with the placenta. They are less likely to breastfeed (which protects babies against infection) and may be at greater risk for depression and post-traumatic stress (which affects how they care for their babies). With all the advances of western medicine, is this really the best we can do?

Ellen Hodnett, a professor of perinatal nursing research at the University of Toronto in Canada argues that we should take a different approach. It turns out, she says, that there is something that reliably reduces pain, distress and the risk of complications and interventions during labour. But it isn't a drug, a scan or a surgical procedure. It isn't a fancy birthing position, or even a state-of-the-art hospital wing. It's having the same caregiver stay with you throughout a birth.

In 2012, Hodnett analysed 22 randomised controlled trials involving over 15,000 women in 16 countries, and found that women who have one-to-one continuous support through labour are less likely to need a c-section or instrumental birth, and are less likely to use painkilling drugs.[5] Their labours are shorter, and their babies are born in better shape. 'It's the only intervention I'm aware of that actually reduces the likelihood of a caesarean birth,' she says.[6]

When used appropriately, c-sections save lives, and all things considered are extremely safe. But they are still major surgery; not something to go through without good reason. The World Health Organization warned in 2010 that although very low c-section rates are dangerous, so are unnecessarily high ones.[7] Studies across different countries suggest that the ideal c-section rate is around

5–10%; rates below 1% and above 15% tend to signify worse outcomes for mothers and babies. The c-section rate in England, where I live, is 26%. In the US it is 33%.[8]

But why should being accompanied by one caregiver – rather than having intermittent support from different midwives who go on and off shift, say – influence whether a woman needs surgery? Hodnett suggests that perhaps those who receive continuous care are more likely to be helped into physical positions that aid labour. Emotional support from a single, trusted person may also reduce women's fear and stress and help them to feel more in control. This can reduce the pain they feel during labour, meaning they need fewer pain-killing drugs, which in itself can reduce complications and cut the need for further interventions. Easing anxiety can also influence the physical progress of labour directly. Hormones released into the bloodstream when we're stressed or scared, particularly in the early stages of labour, act to slow contractions down.[9]

The benefits of continuous care are strongest in developing countries, particularly in situations where women are frightened or uneducated about labour, and tend to give birth in poorly equipped hospitals, without the support of a partner or family member. In a study of 7,000 women across the US and Canada, by contrast, continuous care didn't reduce the rate of interventions at all.[10] Perhaps here medical care is so good they don't need the extra support?

Not so, says Hodnett. She argues instead that an aggressive approach to intervention in these countries trumps any influence of continuous care. 'Everything is ruled by the clock,' she says. 'You've got to have your baby within a certain amount of time or there's a problem. That's not evidence-based, but everybody relies on the clock.' If things don't go to schedule – a labour not starting on time, progressing too slowly, or a woman taking too long to push her baby out – staff step in with drugs, scissors, forceps or surgery.

'You're in an environment in which two thirds of women are getting artificial oxytocin in labour, they're all getting continuous foetal monitoring, so they're confined to bed. They've got IVs, they've got powerful drugs, at least two thirds are getting continuous epidural analgesia in labour.' Women attempting to give birth in such circumstances inevitably end up requiring high rates of drugs and surgery, argues Hodnett, whether they have a supportive caregiver or not.

***

So what happens when women give birth outside of that high-tech environment – for example at home? It's a choice made by about 3% of women in the UK, and just 1% in the US. When women labour at home, the same midwives generally stay with them throughout the birth, while most drugs and medical interventions aren't available without a transfer to hospital.

Randomised trials comparing planned home and hospital births are almost impossible to do, because it's not practical or ethical to force women to give birth in a particular place. But there are plenty of large, observational trials, including a 2011 study that followed nearly 65,000 women with low-risk pregnancies.[11] These studies compare women who choose hospital birth with those who try to deliver at home (regardless of whether they have their babies there or end up transferring to hospital for pain relief or medical intervention). It turns out that simply by choosing home birth, women are less likely to require drugs to induce or speed up labour or relieve pain; less likely to be cut open or to tear; and less likely to need a c-section or instrumental delivery. Their babies are born in better shape and are more likely to breastfeed.

A similar picture comes from UK trials of independent midwives, who work outside the National Health Service. They avoid medical interventions unless there is a clear reason for them, with many of their deliveries taking place at home, and the same midwife cares for a woman throughout her pregnancy, as well as during and after birth. A 2009 study of nearly 9,000 women found that 78% of those in the independent midwife group had unassisted deliveries compared to 54% of those who received conventional care.[12] Their babies were around half as likely to have a low birthweight or to be admitted to intensive care, and breastfeeding rates were much higher.

Perhaps some of these benefits aren't surprising, but aren't the extra interventions carried out during conventional hospital births necessary to save babies' lives when things go wrong? It turns out that in many cases, the answer is no. For low-risk pregnancies in women who have given birth before, labouring at home is just as safe, with exactly the same rate of neonatal death and injury. The authors of a 2012 Cochrane review (the medical profession's gold standard analysis) on home versus hospital births blamed the higher complication rate in hospital on 'impatience and easy access to many medical procedures'.[13] In 2014, the NHS released new guidelines saying that such women are better off outside the obstetrics ward and should be encouraged to give birth either in a midwife-led unit or at home.[14]

It seems that when you replace easy access to technology with caring for a woman's emotional state, she and her baby fare much better – not just mentally but physically too.

***

When I went into labour for the second time, late one October evening, my partner and I called the (independent) midwives and transferred not to hospital but to an inflatable pool on our living room floor.[15]

Jacqui Tomkins arrived first – efficient, expert and the essence of calm. The pain built faster than I expected, every contraction an agonising, all-consuming embrace, each one stronger than the last. And while I had gone into my first delivery naively confident, this time I knew how difficult things might turn out to be. 'I don't think I can do this,' I said to Jacqui. 'Of course you can,' came the no-nonsense reply, like a mother reassuring her child on the first day of school. I'd got to know and trust Jacqui through my pregnancy, so whereas the assurances of constantly rotating midwives during my first delivery only isolated me, this time her words struck home. This was pain but with the fear taken out – nothing like the overwhelming, drowning chaos I'd felt before. Eventually I got into a rhythm: feel it rise, relax, close your eyes, breathe out. Like ducking under a wave into the still water instead of struggling through the crashing surf.

After six hours or so, I heard a noise. It was a guttural roar, that seemed to have come from me. 'What's happening?' I asked in alarm. Jacqui smiled. 'You're pushing your baby out.' This, I discovered, was a different pain, like being ripped and torn from the inside. But it was too late for second thoughts now. And thankfully this final phase is usually short; delivery could be just minutes away. My second midwife arrived, ready for the big moment. Elke Heckel is a large, warm German woman who dresses in bright colours and likes Earl Grey tea. She had heard the noise too. 'Not long now,' she said cosily, and settled herself onto the sofa.

Her arrival was comforting, another thread in the safety net that Jacqui had woven around me. Unfortunately this baby too found an awkward position, with his elbow jammed against his head, and his progress down the birth canal was scrapingly slow. Two hours later, the sun was peeping through the shutters, and London commuters crunched past among autumn leaves. But there was no baby. I was exhausted, and once again starting to panic.

I had been pushing for longer than guidelines for conventional care allow. At this point, NHS midwives would have ambulanced me to hospital for an obstetrician to extract the baby, using scissors, forceps or most likely (because of my previous history) another c-section. This would guarantee a timely delivery. But emergency surgery would bring its own risks, including potential difficulties persuading my newborn to breastfeed. The hospital stay and longer recovery time would also leave me less able to care for my three-year-old at this sensitive time in her life.

Instead, Jacqui and Elke continued to monitor the baby, and assured me that with all looking well, there was no need to intervene. 'You're doing great,' they said. 'He'll come in his own time.' And that was it. That was the moment the statistics changed; the moment an emergency c-section became a complication-free birth. Played out on my living room floor, it was a demonstration of what trials show holds true across tens of thousands of women: that the reassurance of someone we trust is not a trivial luxury. The right words can be powerful enough to replace aggressive medical intervention and transform physical outcomes.

A few minutes later my son slid into the water. Jacqui fished for him in the dim light and guided him into my arms: pale, puffy-eyed, perfect. I was feeding him on the sofa with a mug of tea in my free hand, just in time for my daughter, who had slept through the whole thing, to come downstairs and say hello.

***

Of course, home birth isn't the answer for all – or even most – women. Many women have no wish to give birth at home, and the trials mentioned above suggest that first-time mothers may be safer not to; when they give birth in hospital compared to at home, slightly fewer babies die or are seriously injured. (The same is almost certainly true for high-risk pregnancies such as breech birth or twins, although virtually no studies have been done on this because so few of these women attempt home birth.)[16]

What the contrasting births of my children taught me, however, was how crucial emotional support can be wherever women have their babies. We respond very differently to care delivered by someone we know and trust rather than by a series of strangers, and this affects not just psychological outcomes but physical ones too. Unfortunately, our medical system generally asks women to choose between two extremes: they can either have holistic care at home, but without immediate access to life-saving medical technology; or interventionist, impersonal care in the hospital.

Hodnett argues that we should aspire instead to the best of both worlds: a supportive hospital environment with midwives who stay with women throughout their labour – with access to pain relief and medical technology when they are needed, but *only* then. This is partly the philosophy behind midwife-led birthing centres in the UK, but these still don't guarantee continuous care, and they only cater for women with low-risk pregnancies (about 45% of cases)[17] who are willing to forego the most potent forms of pain relief. What about everyone else though? Wouldn't all women – including those on the obstetrics ward – benefit from more supportive, less aggressive care?

'The common response in North America is we can't afford to have continuous one-to-one support in labour,' says Hodnett. She argues that it doesn't necessarily cost more, however: in a trial of nearly 7,000 women in 13 hospitals across North America she provided continuous care simply by changing how nurses and midwives were deployed, without increasing the number of staff working at any particular time.[18] And of course reducing the number of interventions required would ultimately be cheaper, not more expensive. The average amount charged by US hospitals for maternity care (pregnancy, labour and newborn care) is around $50,000 for women who have a c-section, compared to around $30,000 for those who have a vaginal birth.[19]

If her studies had shown that women should get an expensive new drug during labour, Hodnett says, 'everyone would have gotten it the next day'. Introducing new drugs fits easily into the existing model of medical care. Changing instead how women are looked after wouldn't necessarily be more expensive but it would take wider changes in how hospital departments are organised, and according to Hodnett, there is little appetite for tackling the problem. 'It requires a shift in attitudes and behaviours of the physicians, nurses, midwives and hospital administrators that just hasn't happened.'

In the meantime, women who give birth in hospital continue to receive every medical intervention they need – and many they don't.

***

'Spiderman!' says eight-year-old Daniel excitedly, showing me an oxygen mask that he has decorated with stickers. A ring of tiny spidermen dance around the edge of the mouthpiece. He's sitting on the edge of the bed, swinging his legs, wearing a green hospital robe decorated with stars and hearts.

Daniel is a cute Hispanic boy with dark, spiky hair and a cyst in his brain. Doctors are following the lesion closely to make sure that it doesn't grow, and he's here at Boston Medical Center (BMC) in Massachusetts for his regular MRI scan. 'Spiderman!' he says again. He wrinkles his nose and shows his top teeth in a broad grin.

Like Harborview, BMC serves a challenging demographic. Patients here tend to be poor and disadvantaged. Many of them don't have medical insurance and many don't speak English. When I arrive at eight in the morning on a cold, grey day, the hospital buildings are modern and impressive but the mood is slightly forlorn. Outside, a bulky guy in a Yankees cap announces, 'I'll take you home with me baby,' before asking for some change.

In the cavernous entrance lobby, a teenager dressed in black swears into an iPhone as he circles the spiky palms in giant pots spaced across the floor. Through a door to the left is the radiology waiting room, where bored-looking patients watch a discussion of Kim Kardashian's wedding dress on TV. But as I continue down the corridor, the mood changes. I come to a small but jolly curtained area decorated with kids' drawings and animal photos. There's a pin board covered with cutout kittens. A cupboard packed full with toys. And MRI nurse Pamela Kuzia, smiling and maternal, with pink, flowery shoes.

Kuzia's job is to get the hospital's youngest patients through their MRI scans. This involves lying still in the cramped tunnel of the scanner for around an hour, a daunting experience even for most adults. Where possible, she works to do this without sedating them: 'Our job is not to medicate anyone if we don't have to,' she says. The younger kids and the more anxious ones, she puts under. But even then, getting them into the scanner room and onto the table to be sedated can be a challenge.

Kids like Daniel, for example. He's developmentally delayed. His mother doesn't speak English, and she gets nervous in the hospital, which rubs off on her son. And some of his previous scans have been traumatic, such as when the needle delivering the sedative drugs came out of his vein and fluid started leaking into his arm. It got to the point where Daniel would start crying as soon as he saw Kuzia in the lobby.

But things are different now. When Daniel arrives today, he's calm but quiet, eyes wide. Kuzia hands over some Matchbox cars (she knows they are his favourite), then gives him the oxygen mask to hold. 'This is your pilot's mask,' she says cheerily. 'So, Mr Daniel. Do you like bubblegum or strawberry?' He smiles for the first time. 'Bubblegum!' he says. Kuzia duly sprays the mask and a sickly bubblegum smell fills the space. Daniel holds his newly scented mask proudly, fiddling with the nozzle. By the time he has decorated it with stickers, he's bouncing with excitement.

Then it's time to go to the scanner room. The entrance to the suite could not be more intimidating. It is surrounded by warning signs on the doors, wall and floor. *Stop!* There are red panels. *Danger.* Yellow and black squares. *Caution, magnet always on.* Across the threshold is a large room full of equipment, carried on wheeled stations with screens, wires, buttons and flashing lights. There are spotlights, scissors, gas canisters, drips, pump bags, boxes of gloves and tubes of cream. And in the centre of the floor, humming loudly, a doughnut-shaped tunnel – the scanner itself.

It generates a forcefield tens of thousands of times stronger than the earth's magnetic field, which means that any metal object inadvertently brought into the room – a pen, watch, paperclip or earring – could be propelled towards the scanner (and anyone inside) with deadly speed. It's big and imposing, with a narrow sliding table for the patient that reminds me of the trolleys used to slide corpses in and out of storage freezers in a morgue.

Kuzia has got Daniel to the door. Now she has to get him onto that trolley.

***

You don't need to be a burns or trauma patient, or giving birth, to be confronted with a distressing or painful medical procedure. Millions of people every year undergo not just scans but invasive procedures such as biopsies and keyhole surgery while wide awake. Unlike open surgery, which involves cutting a large incision in the skin, in a keyhole operation the surgeon works through a tiny opening, guided by images from a camera on the end of a tube.

Wounds heal faster than with open surgery and patients can often go home the same day. Patients don't generally require general anaesthesia, and instead receive local anaesthetics combined with sedative drugs. But despite the benefits, being awake while you're operated on can be a daunting prospect. Dangerous side effects limit the amount of sedative drugs that doctors can give safely, and patients typically report high levels of anxiety and pain.

One of the people who carries out such procedures is intervention radiologist Elvira Lang. 'I do surgery on awake people,' she says. 'You have the challenge to get a patient on the table, keep them on the table, and be able to do what you need to do with dignity.'[20] Instead of simply dishing out drugs, she wondered if she could mobilise her patients' psychological resources. So she developed a blend of empathic communication skills, positive suggestion and visual imagery that she hoped would help them to relax, and ease their pain. She calls it Comfort Talk.

While working at Harvard Medical School hospitals in Boston, Massachusetts, Lang tested her approach in randomised controlled trials of more than 700 patients undergoing invasive medical procedures such as breast biopsies or removing a kidney tumour.[21] In the trials she compared her intervention plus standard care ('conscious sedation', where intravenous pain-relieving drugs are available on demand), with standard care alone.

In all of Lang's trials, the patients who received Comfort Talk reported far less pain and anxiety than those who received standard care alone. In a trial of 241 people undergoing renal and vascular procedures, pain scores in the intervention group peaked at 2.5 out of 10 compared to 7.5 for the controls, and their anxiety, instead of steadily rising, dropped to zero.

That's not all, however. Just as in trials of childbirth, Lang found that prioritising patients' psychological state confers hard physical benefits too. Those who received Comfort Talk required much lower levels of sedative drugs, and suffered far fewer complications. In the trial of renal and vascular surgery, for example, patients in the intervention group required only half the amount of drugs. Their procedures were also completed on average 17 minutes faster, saving the hospitals $338 per patient.[22]

But after two decades of work, and the kind of trial results that drug companies would kill for, Lang's ideas were not being taken up by other hospitals. So she decided to disseminate the technique herself and left Harvard to set up her own company, training medical teams in her approach.[23] She's still running clinical trials but now focuses on economic rather than health outcomes 'because frankly that's what hospital administrators are interested in'.

One area she hopes to transform is MRI scans like Daniel's. If patients are too nervous to lie still in the scanner for the hour or so of their exam, the scan has to be aborted – contributing to what is known as the 'claustro rate'. Reducing the claustro rate is a continual struggle, says Kelly Bergeron, manager of the MRI facility at Boston Medical Center. MRI scans can be particularly daunting for patients like those at BMC, she explains, because they tend to be poorly educated and aren't well versed in medicine. 'They don't really know what's happening to them. So to bring them into this sort of technology is frightening.'

If patients fail to complete their scan first time round, they have to come back for a second appointment, generally with the help of a sedative, says Bergeron. But if they are very anxious the drug may not work. 'They fight the medication. What might knock somebody out for a week, this patient is bouncing off the walls.' So they have to come back a third time, perhaps under general anaesthesia, with all the health risks, recovery time and cost that entails.

Lang estimates that such wasted scans cost between $425 million and $1.4 billion each year in the US.[24] If Comfort Talk helps people having biopsies and keyhole surgery, could it get them through MRI too?

***

'Stinging coming up!' 'Another sharp jab in a minute.' 'You're going to feel some burning.'

Warning patients about pain or discomfort they are about to feel is a staple of conventional medical care. But Lang argues that during medical procedures such as scans or operations, we are particularly susceptible to the nocebo effect, and that being told how much things are about to hurt simply worsens our pain. 'As soon as you put your foot into a medical facility or a dentist's surgery, you are already in a hypnotic state,' argues Lang. 'You are very highly suggestible.'

To prove her point, Lang worked with Harvard placebo researcher Ted Kaptchuk to analyse 159 videos of patients going through surgery as they rated their pain and anxiety every 15 minutes.[25] In the videos, medical staff often gave graphic warnings about upcoming pain (including the lines listed above). When patients received these warnings ahead of potentially painful events such as injections or puncturing the skin – even if the negative words were preceded by qualifiers, such as 'not much' or 'a little' – their pain and anxiety scores soared.

A key part of Lang's Comfort Talk approach, then, is eliminating negative or scary language. Instead of continually telling patients how much pain they're going to be in, Lang suggests warning of any negative possibilities beforehand, during the informed consent stage. 'But after all that is done and the patient is there, you don't need to say that it's going to sting or burn.'

Lang recently trained Bergeron's team in the use of Comfort Talk. Although Bergeron thinks her team already dealt with patients sensitively, she says Lang helped them to rethink the language they use during scans. Before their training, her staff routinely warned patients of upcoming discomfort, for example when they were about to be injected with contrast dye, which helps certain tissues to show up more clearly on the scan. 'Now there's no mention of a needle or a bee sting,' says Bergeron. 'We took all of that out. Now we say, "I'm going to give you the contrast."' Instead of locking patients into the scanner, staff make them comfortable. The panic button has become the call bell.

Another component of Lang's approach is encouraging patients to visualise positive images. One of the scariest parts of an MRI scan is being immobilised in the 'head coil', a plastic mask that locks in place over your face. Bergeron and her colleagues now suggest to children that they are flying in a rocket ship, or that the head coil is a football mask. For adults, they might suggest lying back on a massage table, and even provide a choice of orange or lavender aromatherapy tabs to supplement the illusion of a trip to the spa.

And if a patient is very nervous, staff can read from a script.[26] This is presented as a relaxation exercise but is similar to a hypnotic induction, in which patients are invited to roll their eyes upwards, breathe deeply and focus on a sensation of floating, before imagining a pleasant setting of their choice. Bergeron says her team felt awkward doing this at first, but soon saw the benefits. 'You can read from that piece of paper and it still works,' she says. 'If they are listening, it'll calm them down. It's kind of crazy but it does really work.'

In a study of around 14,000 MRI appointments, Lang showed that training MRI teams to use Comfort Talk reduced the claustro rate by nearly 40% (saving hospitals $750 to $5,000 for each saved scan, depending on the insurance carrier and facility).[27] She has found similar results in a so-far unpublished study of 90,000 appointments, in which the BMC team participated.

Despite Lang's positive results, Bergeron predicts a struggle persuading the medical establishment to embrace Comfort Talk. 'Because it is really not medicine at all, it is just a mindset,' she says. 'It is very difficult to bring that sort of tool or mentality into a western-based health system that is driven by testing and results.' Nonetheless, she says that since her team started using the approach, patients are more likely to get through their exams, scans go quicker with fewer interruptions, and fewer patients need to be sedated with drugs.

Even better, 'I haven't seen a screaming kid go in there in I can't tell you how long,' says Bergeron. 'It's been one of the biggest benefits of this whole thing.'

***

Kuzia takes Daniel across the danger threshold into the scanner room. She walks him around the beige tunnel, watched by white-coated medical staff though a window from the control room next door. 'This is the big room. Inside here is the big giant camera.'

She encourages Daniel to slap the side of the tunnel, then points to the table. 'Come and sit right up here.' He hops up and she keeps talking. 'Here's your pilot mask. All spaceships need a pilot mask. How about some stickers on your belly? One, two, three, four.' He sits quietly as she sticks four ECG pads onto his chest to monitor his heart, and slips a blood pressure cuff on his arm.

The anaesthetist stretches out some plastic tubing and attaches it to Daniel's mask. 'Pilots have oxygen when they're high in the sky,' he says. 'Your job is to breathe the oxygen.' Daniel holds the mask to his face and inhales. Then he cries out, and Kuzia holds him close. 'Think about a Spiderman adventure,' she whispers. Then two seconds later, 'He's asleep.' Daniel has done perfectly.

Next time, says Kuzia, he can try his scan without sedation. Simple changes such as avoiding scary language, letting him choose a scent and decorate his mask with stickers, and encouraging him to imagine a spaceship ride, have transformed a screaming, resistant boy who had to be sedated in traumatic circumstances into an engaged one who from now on hopefully won't need any drugs at all.

Kuzia lays Daniel down and covers him with a sheet, before sliding him inside the scanner. He's surrounded by beeps and pings, and his heart rate pulses green on a nearby monitor. Next door, in the control room, slices of his brain appear on the computer screen, floating white on black.

***

Across town at Massachusetts General Hospital, Vicki Jackson cares for people with terminal illness. As a palliative care specialist, her job is not to prescribe drugs or treatments, but to talk. She confronts questions that people facing death often aren't asked: how much do they want to know about their prognosis; would they rather reduce symptoms or lengthen life; where and how do they wish to die? Jackson's primary aim is to increase quality of life during the time that these patients have left. But in a pioneering trial published in 2010, she found that these discussions can do far more than that.

The study, led by oncologist Jennifer Temel, followed 150 patients who had just been diagnosed with terminal lung cancer.[28] Once diagnosed, these patients typically have less than a year to live. Half of the patients in Temel's study received standard cancer care. The doctors were focused, as you might expect, on the patients' medical condition: planning their treatments, monitoring progression of their tumours and managing any complications. The rest of the patients received exactly the same treatment but were also offered monthly sessions of palliative care.

During these sessions, Jackson and her colleagues focused not on the medical details of the patients' cancer but on their personal lives, including issues such as how they and their family were coping with the diagnosis, and with any side effects of their treatment. For example, Jackson tells me about a patient with pancreatic cancer – let's call him Peter – whom she saw the day before our interview, after his latest scan revealed bad news.

'His oncologist spent 40 minutes going over the scan results, and then I spent another hour going over it with him,' she says.[29] The message from the oncologist was that Peter is unlikely to benefit from further chemotherapy; Jackson's task was to discuss with Peter what that means in terms of how he should live his life. 'His son is getting married in six months. I don't think he is going to make it to the wedding,' she says. 'How is he going to talk to his children, who live all over the country, especially his son?'

Jackson says she couldn't do her job without getting to know her patients as rounded people – their interests, values and families. Good palliative care isn't so much about helping people to die as helping them to live, she says. Working out how to do that requires figuring out who they are as a person and what living means to them, whether it's playing golf, watching soap operas, or being well enough to attend a wedding. 'For every person it's different.'

On average, the lung cancer patients in Temel and Jackson's study received four sessions of palliative care. The results were striking. Compared to a control group, these patients had much better quality of life (a measure that includes ratings of physical symptoms) and felt significantly less depressed. They also received much less aggressive care at the end of their lives, with fewer rounds of chemotherapy and longer hospice stays. But the researchers were surprised to find something else. The palliative care group survived for an average of 11.6 months, compared to 8.9 months for the control group.[30]

It'll take more and larger studies to confirm this result and pin down exactly why simply talking to a palliative care specialist had such a dramatic effect. The lower rates of depression may be one factor – in general, cancer patients who are depressed don't live as long. It's probably also because aggressive treatments given at the end of life, when patients are very sick, can hasten death rather than delay it.

When patients had the opportunity to talk to someone not about their tumour but about what they wanted from the time they had left, they made different choices. They still chose aggressive care early on, but in their final few months switched their focus to maximising quality of life. They received less last-ditch treatment and, alongside all the other benefits, seem to have survived longer as a result.

By contrast, in the standard model of care, argues Jackson, aggressive treatments are the only thing on offer. People with terminal cancer accept round after round of chemotherapy because in the absence of any alternative, not doing so basically means giving up.

'Intervention becomes synonymous with hope,' says Jackson. 'And it's not.'

***

All too often when we receive medical treatment, our mental state is seen as a secondary concern, and our role as a patient doesn't go much beyond signing consent forms and requesting pain-relieving drugs. When I gave birth to my first child, I received state-of-the-art medical care, but I felt (as many women do) like an object on a conveyor belt, a passive recipient of a bewildering series of medical interventions that started with breaking my waters and ended in emergency surgery. We often focus on the importance of pain relief during childbirth, but I ultimately found that loss of control more distressing than the physical pain I felt when later giving birth without powerful drugs.

The three projects described in this chapter – midwives supporting women during childbirth; radiologists changing how they talk to patients; and doctors discussing difficult questions with the terminally ill – instead give patients an active role to play. These might seem like common-sense interventions, but they all embody a fundamental (and for our medical system, revolutionary) shift in what it means to care for someone. Medicine becomes not an all-powerful doctor dishing out treatments to a passive recipient, but a partnership between equal human beings.

This principle is at the heart of many of the other cases we've seen so far too, including the IBS patients treated in Peter Whorwell's hypnotherapy clinic, Manfred Schedlowski's kidney transplant recipients, and the burns patients immersed in Hunter Hoffman's Snow World. Instead of medicating their way out of problems with ever-greater doses of drugs and interventions, these medical professionals are harnessing their patients' psychological resources as a critical component of their care. They're doing this for adults and children; for chronic complaints and for emergencies; from birth until death.

This approach provides a better experience for patients. It costs less. And it improves *physical outcomes*. Patients suffer fewer complications, recover faster, and live longer. The trial results show that individual cases like Daniel's and mine aren't lucky coincidences, but reflect a wider picture that holds strong across hundreds of thousands of patients. We are humans, not machines, after all. When we're receiving medical care, our mental state matters. Those who feel alone and afraid do not fare as well as those who feel supported, safe and in control.

What, then, about the rest of the time? We spend the majority of our lives not as patients but as people in the push and pull of everyday crises – dealing with tricky relationships, stressful jobs and bad traffic; negotiating deadlines, disappointments and debt. In the second half of this book, we'll look beyond medical therapies and treatments to investigate the importance of our minds from day to day. How do our thoughts, beliefs and emotions influence our physical health over the course of our lives?

8

FIGHT OR FLIGHT
Thoughts That Kill

At 4.30 a.m. on 17 January 1994, Los Angeles was hit by a devastating earthquake. With a magnitude of 6.7, it was the most powerful quake that had ever struck a major US city. Shockwaves generated 11 miles below ground ripped through the city for ten terrifying seconds. Apartment buildings collapsed, bridges and power lines toppled, hospitals were wrecked, and a 64-car freight train was derailed. Dozens of people were killed and thousands injured as the lights went out across the city and fires raged out of control.

Robert Kloner, a cardiologist working at the Good Samaritan Hospital in downtown LA, was asleep at his home when the tremors started. 'The lights went out, and the house shook like a train,' he recalls. 'Everything made of glass broke, our windows cracked, the wall of the bedroom partly came down.' As alarm flooded Kloner's body, his heart raced and his blood pressure spiked. 'It is one of the few times in my life that I felt I might die.'[1]

Few demonstrations of the mind's effects on the body are as dramatic as sheer terror. Kloner was lucky enough to survive the quake unscathed. But he later discovered that for dozens of others who lived in the area, simply thinking that they were about to die was enough to kill them. The official death toll from the LA quake was 57, including people caught in the rubble of their homes, and a police officer on his motorcycle who fell 40 feet to his death when a freeway collapsed. But when Kloner studied cardiac deaths reported across the county during the period preceding the disaster and on the day itself, he uncovered a group of hidden victims.[2]

For the two weeks before the earthquake, an average of 73 people each day died of heart attacks. But on this terrifying day, the number jumped to 125, far beyond the usual range of variation. This suggests that around 50 people's hearts failed as a direct consequence of the disaster. Spikes in cardiac deaths have been seen for other crises too,[3] for example during the Iraqi missile attack on Israel in 1991 and the devastating earthquakes that hit Athens, Greece, in 1981 and Kobe, Japan, in 2005. Rather than being crushed by falling masonry, the extra victims were literally scared to death.

***

If you have ever stepped out in front of a car, or been woken by a scary noise in the dead of night, you'll know how violently your body can respond to fear. Within a split second of sensing a threat, you feel a jolt of adrenaline as your heart beats faster, you breathe more heavily and your pupils dilate. Blood is diverted away from non-urgent areas such as the gut and sexual organs and towards the limbs and brain. Digestion slows, while fat and glucose are released into the bloodstream to fuel your next move.

This emergency response is known, of course, as 'fight or flight'. It's controlled by stress hormones released into the bloodstream, including adrenaline and cortisol, as well as the sympathetic nervous system, which connects the brain to the body's major organ systems (and is behind the conditioned responses described in Chapter 4).

Fight-or-flight evolved initially as a response to physical trauma or stress: injury, exhaustion or starvation. But it can be triggered by psychological factors too. No need to wait until a predator bites. Our body is on alert as soon as we see, smell, hear – or even imagine – a threat.

The bursting blood pressure and racing pulse triggered by the perception of danger are at times so severe they can kill us, as Kloner found. Of course, dropping dead from fright is an extreme phenomenon that affects a relatively small number of people. Kloner tells me that it is most likely to affect those who already have weak hearts, and requires an intense situation in which you feel 'personally, physically threatened'.[4] In general, the fight-or-flight response is helpful: an instinctive reaction that has kept our ancestors alive in fast-changing environments over millions of years of evolution. It switches on in a heartbeat, and when the threat is over, our bodies relax again.

Or that's how it works in most species. As Robert Sapolsky, a pioneering stress researcher at Stanford University, describes in his 1994 book *Why Zebras Don't Get Ulcers*, a zebra being chased by a lion benefits from the full force of its fight-or-flight response. When the chase is done, the zebra recovers (assuming it hasn't been eaten) and its physiology returns to normal – the picture of rest and calm. The animal doesn't replay the twists and turns of the chase in its mind, or mull over whether it will be as lucky the next time.

But people are different from zebras. Our more sophisticated brains have given us the ability to learn from our mistakes and to plan for the future – but also to worry about our problems all the time. From terrorist attacks, redundancy or relationships to bad traffic or a quarrel with a friend, we replay past situations and agonise about future ones. We call this stress, and it triggers the same emergency response in the body as being caught in an earthquake, albeit to a lesser extent. We could be sitting by the fireplace at home, surrounded by friends and eating a hearty meal, yet our minds and bodies are still on high alert.

Fortunately, these day-to-day concerns don't strike us down on the spot. But given time, they can be just as deadly.

***

Lisa's life is bound by laws that she can't predict or understand. 'I live in fear of breaking one of Brandon's rules,' she says. It could be a slight change to the daily routine, a step or movement out of place, or something that she has no control over at all. 'Sometimes I don't even know what's going to set him off, and then he's going to cry and scream. He can turn into an animal when he gets upset.'

Lisa is a 42-year-old economist from San Francisco, and Brandon is her son. Four years ago, he was diagnosed with high-functioning autism. Caring for him is a challenge every minute of the day, so I've called her to find out what it's like to live with constant, unrelenting stress.

At first, says Lisa, she just thought her toddler had a quirky, quiet personality. But as Brandon got older, it became clear something was wrong. He would repeat words, or open and close doors nonstop, for perhaps 20 minutes at a time. After the diagnosis of autism, the family took a different course. Lisa gave up her full-time job (she now works part-time) to look after Brandon and his older brother Nathan. But Brandon's behaviour continued to deteriorate. He was immersed in his own imaginary world and would lash out with violent tantrums.

Brandon is now eight years old. I ask Lisa for a photo and she emails me one taken at home earlier that day. Mother and son are sitting together on the floor, leaning against a sofa, relaxed and smiling. Brandon looks adorable in a blue t-shirt, with light brown hair and a cute grin directed straight at his mum.

The shot looks carefree but as I listen to Lisa's story, I realise that it has taken her years of hard work and heartache to get to this point. For a year or so, Brandon's behaviour was so bad that

Lisa couldn't leave the house. 'I thought he was going to end up in an institution,' she says. After getting help from a behavioural therapist, life has become more manageable. Lisa now does play therapy with her son every day, encouraging him to interact and to make eye contact. He's obsessed with maps, she says, and has memorised the whole San Francisco public transportation system. 'If I'm playing along with his pretend world, he's really delightful.'

'But I have to continually stay on it,' she adds. 'It's not like I can relax.' Brandon goes to a normal school but is falling behind academically, she says, and doesn't have any friends. At recess, when the other kids play together, he walks around the edge of the playground pretending that he's a bus driver. Lisa is convinced that he wants to interact, but doesn't know how.

'It's heart-breaking,' she says. 'If he sees someone hurt on the playground he goes over to them and wants to help. But he doesn't know what to say.' Brandon needs a one-to-one aide at school, which he hates, and which acts as a further barrier between him and the other children, so she's looking for a school where he can be more independent. 'I'm devoting my life to getting him into the right environment.'

At home, Lisa lives her life in 15-minute intervals. 'I have to constantly either give him something to do or interact with him directly otherwise he's going to have trouble,' she explains. 'From the moment I wake up, I have to plan the day, how it's going to go. And then hope for the best.' The worst times are when Brandon gets upset, which is a lot. He cries and screams, sometimes for hours. 'One time he came out of a church group and it hadn't gone well,' says Lisa. 'He punched me in the stomach. I was just like wow, I can't hand him back. I'm going to have to be Mother Teresa just to love him.'

What kind of things upset Brandon, I ask. He gets overwhelmed by stimulation, she replies; the sound of laughter if they have visitors, for example. 'He'll start screaming because it's so loud.' Small details that don't go his way can also derail him. That includes any change to the daily routine, like when she picked Brandon up from school and Nathan wasn't in the car as usual because he'd had a doctor's appointment. When his brother steps on one of his maps. Or the time she ripped a piece of paper to write something down.

'Oh gosh,' she says. 'He didn't like that I ripped the paper, he threw a fit about that.'

There's a pause, and I realise that Lisa has been speaking through her tears. I try to imagine what it must be like. The exhaustion, the uncertainty for the future. The unpredictability and the struggle to connect. The desperation of having a child imprisoned, alone and frustrated, in a world that you can't rescue them from; a world that you can only ever glimpse.

I'm sorry, I say, and I don't just mean for making her cry.

***

The challenges of caring for Brandon have often pushed Lisa to breaking point. 'When he is freaking out, I hate to admit it, but sometimes I would freak out too,' she confesses. And they have pushed her family past it. She and her husband are in the process of separating. They remain on good terms and plan to set up two caring homes for their children, but their bond has broken under the strain of their son's condition. 'I can't deal with my husband and the kids,' says Lisa. 'It's one or the other.' The devastating psychological and emotional effects of her situation are clear. But what about the physical impact?

Over the last few decades, scientists have realised that constant stress can ravage our bodies. Not surprisingly, the cardiovascular system is particularly susceptible. Switched on long term, the raised blood pressure triggered by the fight-or-flight response can damage blood vessel walls, eventually causing problems from clogged arteries to heart attacks. Trials that followed tens of thousands of British government workers – known as the 'Whitehall

studies' after the London street on which the government buildings are located – have found that those with more stressful jobs
die significantly younger, mostly because of heart disease.[5] In
Eastern Europe amid social collapse after the fall of Communism,
death rates from heart failure spiked.[6]

Chronic stress reaches beyond the heart, however. During fight-
or-flight, the body burns fuel to raise blood sugar levels. This
gives us a crucial energy boost, but over time it can increase the
risk of obesity and diabetes. And it plays havoc with our immune
system.

Until a few decades ago, scientists didn't think it was possible
for psychological stress to affect the body's response to infection,
but there is now a flood of evidence proving the link. The effects
are complex, but in general, acute bursts of stress (lasting minutes
to hours) seem to boost the immune system in readiness for
injury, an effect that's mediated by stress hormones including
cortisol.[7]

Once the stressful event is over, levels of these hormones quickly
return to normal; cortisol acts as its own off-switch, for example.
It's a clever system that ensures the activated immune cells – which
cost energy, and might attack the body if switched on too long
– are only around as long as they are needed.

When we're under chronic stress, however, cortisol is released
into the body all the time. This acts as a permanent off-switch,
and suppresses the immune system. Chronic stress impairs our
response to vaccines, and makes us more susceptible to infections
from the common cold to HIV.[8]

And if we're too stressed for too long, the off-switch can wear
out, and our bodies no longer respond to cortisol as they should.[9]
This allows the immune system to rage out of control, leaving us
more susceptible to allergies, and most damaging of all, chronic
inflammation. Visible as the swelling and redness that appears
around a scratch, inflammation is the body's first line of defence
against infection and injury. Tiny blood vessels dilate and become

leaky, allowing blood and immune cells to spill out into the surrounding tissue. This can clear an area of irritants, invaders and damaged cells quickly and effectively, and a brief spike of inflammation is a crucial part of wound healing.

But switched on long term, too much inflammation disrupts the process and wounds actually mend more slowly – researchers have seen this in women caring for a relative with Alzheimer's disease, in dental students facing exams, and in married couples when they fight.[10] High levels of inflammation exacerbate auto-immune diseases from eczema to multiple sclerosis. And over time, inflammation eats away at healthy tissues such as bones, joints, muscles and blood vessels; one stress researcher I spoke to calls it 'the juice of death'. In Europe and the US, around a third of us have inflammation levels that are dangerously high,[11] and scientists are realising that this causes or contributes to conditions including diabetes, heart disease, arthritis, osteoporosis and dementia – all of the chronic diseases that plague us as we age.[12]

The physiological changes caused by stress seem to play a role in some cancers too. Many epidemiological studies, following millions of people over time, have found that even after controlling for behavioural factors such as smoking and drinking, stressful life events increase the risk of certain cancers. (Others don't see an effect, however, possibly because any link is likely to depend on the type of stress, the body tissue affected, and the developmental stage of the tumour.)[13] Meanwhile lab experiments suggest that stress inhibits DNA repair mechanisms, at least in animals, and that it suppresses parts of the immune response, such as natural killer cells, that normally fight tumours.[14]

And by boosting inflammation, which clears away damaged cells and promotes the growth of new blood vessels, the fight-or-flight response provides just what a developing tumour needs: a local blood supply and space to grow. If mice with various cancers are subjected to stress, or injected with the stress hormone adrenaline, their tumours grow and spread faster.[15] (Giving them a drug that stops adrenaline binding to cells blocks the effect, and several research groups are now studying whether similar drugs in humans – called beta-blockers, already widely used in healthcare to treat hypertension – have a similar protective effect.[16])

As if all this wasn't enough, there is one more problem that stress can cause, arguably the worst of all. In 2004, Elissa Epel and Elizabeth Blackburn at the University of California, San Francisco, measured the effects of stress on repeating DNA motifs on the end of chromosomes, called telomeres, which play a crucial role in the ageing process.[17] These caps shield the ends of our chromosomes each time the DNA is copied and our cells divide. But they are worn down themselves in the process. When telomeres get too short, cells malfunction and lose their ability to divide, meaning that our tissues are no longer able to renew themselves.

Epel and Blackburn studied telomeres in two groups of mothers: one with healthy children; and one that, like Lisa, had children with chronic conditions such as autism. It turned out that the more stressed the women felt, the shorter their telomeres were.[18] The most frazzled women had telomeres that looked ten years older than those of the women who felt least stressed, and their levels of telomerase, an enzyme that rebuilds telomeres, were halved. In other words, the researchers claimed, feeling stressed doesn't just make us ill. It ages us.

The study was described by stress researcher Robert Sapolsky as 'a leap across a vast interdisciplinary canyon',[19] connecting the women's complex lives and experiences to the molecules inside their cells. Many telomere experts were sceptical at first but Epel and Blackburn's paper sparked a burst of research, and stress has now been linked to shorter telomeres in many different groups, including older women; Alzheimer's caregivers; victims of domestic abuse, rape and early life trauma; and people with psychiatric disorders such as depression and PTSD.[20]

'Ten years on, there's no question in my mind that the environ-ment has some consequence on telomere length,' says Mary Armanios, who studies telomere disorders at Johns Hopkins School of Medicine in Baltimore, Maryland.[21]

People with shorter telomeres are more likely to have stress-related conditions such as diabetes, heart disease, Alzheimer's and stroke, and they die younger.[22] The big question for researchers now is whether short telomeres directly contribute to illness and mortality, or are just a harmless side effect of age-related damage. Seriously damaged telomeres clearly devastate health: people with the genetic disorders studied by Armanios, who have much shorter telomeres than normal, suffer from accelerated ageing and organ failure.[23] But she questions whether the smaller changes caused by stress will turn out to be signifi-cant, especially because telomere length is quite variable in the first place.

Blackburn, on the other hand, says she's increasingly convinced that psychological factors are important. Genetic mutations that shorten telomeres to a lesser extent than the extreme cases studied by Armanios still increase future risk of a range of chronic condi-tions, she points out.[24] And variations in telomere length equiva-lent to those caused by stress seem to predict future health even after taking into account traditional risk factors such as body mass index or blood sugar levels.[25]

The link with ageing comes as no surprise to Lisa. Four years on from her son's autism diagnosis, I ask whether her stress has had a physical impact. Yes, she says. She's 42, with hair that's naturally light brown, like Brandon's. 'But all of a sudden, over the past three years, my hair has turned grey.'

***

I drive east and then south from Atlanta, Georgia, until the city is long gone and the sun slants through the pine trees, making zebra stripes across the tarmac. Tom Petty is on the radio and birds of prey hover above, eyeing up the abundant roadkill.

After a couple of hours I reach the outskirts of a town called Milledgeville. The roads disintegrate into narrow lanes with rough edges, and the whole place feels forgotten. There are battered wooden homes behind wire fences, and some trailers with plastic chairs out front. At one point, my car's navigation system directs me down a dead end; the tarmac falls into gravel that dissipates among the trees, in front of a stained white house with tiny windows and wooden legs.

Milledgeville is located in a crescent-shaped strip of land across the south-eastern United States that's informally known as the 'black belt'. In the nineteenth century, this name was inspired by the colour of the region's unusually fertile soil, which was home to cotton plantations worked by slaves. It later came to refer to the high proportion of African Americans who live here, typically above 50% of the population.

Many people here now suffer debilitating poverty. The black belt traverses jut 11 states, which are home to around a third of the nation's poor. The region is characterised by substandard housing, schools and transport, as well as high crime and unemployment, all problems that disproportionately affect African American residents.[26]

And according to Gene Brody, a psychologist at the University of Georgia who studies the health of black belt families, these families have some of the highest levels of chronic disease in the country, including heart disease, diabetes, stroke and cancer. Stress, it turns out, doesn't just affect individuals. In places like Milledgeville, it blights the health of entire populations.[27]

I'm interested to know what life is like here, so Brody has put me in touch with some of the residents, including Susan. When I eventually reach her home, I find a sturdy brick bungalow – the nicest house on the street – with brick steps leading up to the front, and a brick patio at the back. Bluebirds and red cardinals flash past. A sprawling yard, home to a battered pick-up truck and piles more bricks, backs straight into the woods. Coyotes are regular visitors here, Susan tells me later, along with foxes, rabbits and wild turkeys.

She opens the door with an excited white dog in her arms. 'We're in the middle of a house clear,' she apologises, as she ushers me past a cluttered hall into an immaculate living room. On one wall is a huge ornate mirror, on another, two miniature gold violins. There's a fluffy turquoise rug and cushions with long fringes, while the shelves are filled with family photos and cut glass.

Susan has short, grey hair and no make up, and is casually dressed in bright pink jogging bottoms and a baggy Georgia College Bobcats t-shirt. When she greets me, her voice is strong and resonant.

She was brought up in Milledgeville, she says, in a 'shotgun house' – so-called because you could stand and look straight through – with an outdoor toilet, and two outdoor water taps shared between nine families. 'We had a huge black pot to warm the water,' she recalls. They made their own soap, and hog head cheese from fragments of pork. She lived with her grandparents. ('I knew who my parents were,' she says. 'But they were just somebody I knew.') Her grandfather spoiled her, but her grandmother kept her in line with a switch. Lots of her friends missed school to pick cotton – Susan wanted to but her grandmother wouldn't let her. 'She said it eats all the skin from around your nails, and ruins your hands.'

Susan is now a central member of this community; she's active in the church, and volunteers at a local children's centre. She has been married to her husband, George, for 50 years. It's clear they have fought hard for what they have. George built this house himself, she says, using bricks scavenged from other houses that were being demolished. She points out the huge fireplace, made of bricks from her childhood home.

When I ask what life in Milledgeville is like today, she tells me that unemployment is a huge problem. The farm jobs are long gone, and she has seen most of the area's big employers disappear too: Mohawk, which makes carpet fibres; manufacturing company JP Stevens; Oconee Brick. Young people here, she says, have largely given up. 'They're not prepared to go to college,' she says. 'All they want is easy money.' Instead, the area 'is corrupted with drugs'.

The challenges of life here show up in official statistics. More than half of African American children in the rural South live in poverty, and most of these grow up in single-parent households. Life on a low income can be even harder here than in inner city neighbourhoods, says Brody: it's impossible to get around without a car, there are no jobs, and there's nothing for young people to do. Youth drinking (and its consequences, such as failure at school, bad behaviour and risky sex) is rising faster in rural areas than in cities, with rural black teens now drinking as much as or more than their urban peers.[28]

Susan has four children, all now grown. She aimed to instil Christian values in them, she says, along with respect for their elders. But that wasn't enough to save her daughter, Jennifer, from the lure of drugs and crime. Susan recalls chatting to a neighbour on the phone one day when the operator cut in and told her to go immediately to the police department's public safety building. Jennifer's 16-month-old daughter Jessica had just been rescued from a house where she and another child of similar age had been left alone for the entire day – completely ignored by three men who were apparently out on the porch.

Jennifer, it turned out, was in another county, in jail. More than 20 years later, Susan still remembers walking in and seeing her granddaughter and the other child sitting on the floor, with food on Styrofoam plates between their legs. She and her husband took Jessica in that day. They were already looking after her older brother Kevin, and eventually cared for Jennifer's third child too.

Jennifer has caused her lots of stress over the years, says Susan, like when she'd turn up demanding the kids. One time she disappeared with Kevin for several days; Susan and her husband were worried sick until they finally tracked them down in a motel. Now the grandchildren have left home, though, she rarely sees her daughter. 'What do we need her for now? They've grown.'

But Kevin still causes her heartache. After a brief spell in the military he left the army, came back to Milledgeville and got in with a bad crowd. A few weeks before my visit, he got out of jail and turned up at home, Susan says, wanting to move in. She told him to leave: 'I can't live with people who are going to steal from me.'

***

Susan felt so ill after the confrontation with her grandson that she went to the doctor, who put her on medication for high blood pressure. Indeed, living in this type of neighbourhood – characterised by high crime, drug use, single parenthood, no job opportunities – can have dire consequences for health across the life span. Children in low-income families are more likely to be born small, born early and to die shortly after birth. As they grow up they have more health problems, including obesity, insulin resistance and asthma. Later in life, they are more likely to be ill and to die from stroke, cardiovascular disease, chronic lung disease and some cancers.[29]

The difference in health between rich and poor varies in different countries, roughly correlating with the level of economic inequality in that country.[30] According to Greg Miller, a psychologist from Northwestern University in Evanston, Illinois, who studies the effects of poverty on health, it's much more dramatic in the US than in Canada or Sweden, for example, with Britain somewhere in between. 'But disparities in health are persistent across almost all countries we know of, regardless of whether they are modern, industrialised countries or developing countries,' says Miller. 'You see them within countries, you see them between countries, you see them in women and in men, within different ethnic groups, and at every stage of the lifespan all the way from pregnancy outcomes to dementia and stroke.'[31]

What causes the difference? The effect isn't explained by access to healthcare or material resources; if this were the whole story, everyone above a certain threshold of basic need should have similar health. Instead, there's a linear health gradient through the entire socioeconomic spectrum, right up to the most privileged groups. And although people living in poverty tend to lead unhealthier lifestyles (for example drinking and smoking more and exercising less), when researchers account for this, the effects on health don't disappear. In addition to behavioural factors, argues Miller, the stress and alienation of being poor cause chronic inflammation that damages health throughout people's lives.

In particular, the environment we are exposed to as children seems to influence our susceptibility to stress later on. For example, some children from poor families work hard, go to college and leave for good jobs elsewhere, where they lead lifestyles just like those of their more privileged peers. They have low rates of drug use and behavioural problems, and appear to be perfectly healthy, says Brody. 'But if you unzip the exterior and look at their biology, they look different.' They have higher blood pressure, and higher levels of circulatory stress hormones and inflammation.[32]

Regardless of their current circumstances, those who grow up in adverse environments also have increased rates of cancer, heart disease, illness and death from all causes. One study followed more than 12,000 Danish adoptees and found that mortality in their forties depended on the social class of their biological father, but not their adoptive father.[33] Another tracked medical students enrolled at Johns Hopkins University, for 40 years.[34] Among these educated, well-off doctors, those who had grown up in poor families were more than twice as likely to suffer from heart disease by the time they were 50.

Stress from adversity and inequality also seems to be a major force eroding telomeres. People who didn't finish high school or are in an abusive relationship have shorter telomeres, for example, while studies have also shown links between short telomeres and low socioeconomic status, shift work, dangerous neighbourhoods and environmental pollution.[35] In African Americans, experiences of racial discrimination have been associated with various biological markers of stress, including shorter telomeres.[36]

Again, children are particularly at risk. Being abused or experiencing adversity early on – including in the womb, through exposure to a mother's stress hormones – leaves people with shorter telomeres for the rest of their lives.

Results like this are leading some scientists to argue that to tackle the rising epidemic of chronic disease, governments need to reduce social inequality and in particular to support women of childbearing age. In 2012, Elizabeth Blackburn and Elissa Epel wrote a commentary in the prestigious science journal *Nature*, calling on politicians to prioritise 'societal stress reduction'.[37] The stress that women experience during pregnancy and while raising their children causes health problems and economic costs in the next generation for decades to come, they pointed out, even if those children later escape to more comfortable circumstances.

There's now compelling evidence that the biology of how we age is shaped early in life, Epel tells me. 'If we ignore that, and we just keep trying to put band-aids on later, we're never going to get at prevention and we're only going to fail at cure.'[38]

***

Tackling social inequality is hardly a new idea, yet the sheer scale of the health problems caused by stress and poverty – and the finding that the circumstances we grow up in shape our disease risk for life – arguably provide a stronger case then ever for governments to act. But perhaps politicians still aren't ready to leap across the interdisciplinary canyon that Blackburn and Epel bridged a decade ago. According to Epel there was little response to the *Nature* article. 'It's a strong statement so I would have thought that people would have criticised it or supported it,' she says. 'Either way!'[39]

There are some efforts to put her vision into practice, however, and in Chapter 10 we'll look at what happens when researchers try to buffer the effects of stress in some of the communities that need it most, including Milledgeville. But in the meantime, is there anything we can do as individuals to protect ourselves from the debilitating effects of stress?

Few of us can remove all the stress from our lives, any more than Susan can change her neighbourhood or Lisa can hand Brandon back. But there is some good news. External problems – debt, rocky relationships, having a child with autism – do not generally damage our bodies directly. What harms us is our psychological response to those circumstances; not the state of our environment, but of our *mind*. And that is something we can control.

Wendy Mendes, a psychologist at the University of California, San Francisco, uses the example of a skier who unexpectedly comes across a steep, icy trail; it's her only way down the mountain. The skier's heart rate will likely rise as her body prepares for the descent. But depending on how experienced she is and whether she believes she has the skills to cope, her predominant emotion could be either fear or exhilaration.[40]

These contrasting mental states are both versions of fight-or-flight, but they have very different physical effects on the body, says Mendes.[41] Both scenarios trigger the sympathetic nervous system (SNS), but excitement or exhilaration activates it to a greater extent. From an evolutionary point of view, this is the mindset of a hunter closing in for the kill; a runner being chased but confident of escape; a fighter who knows he has the upper hand. Our peripheral blood vessels dilate and our heart works more efficiently, pumping oxygenated blood to the limbs and brain. People experiencing this type of response perform better than normal, not just physically but mentally too.

Fear, on the other hand, causes the body to go into damage-control mode as it prepares for defeat. We're being hunted and there is no escape. We're in a fight against a stronger adversary. The SNS is activated but to a lesser extent. Our peripheral blood vessels constrict and our heart beats less efficiently, so less blood is being pumped around the body. This serves to minimise blood loss if we are caught and injured. But it impairs our performance and strains the cardiovascular system, because the heart is forced to work harder to push blood around the body. In addition, there's a surge of the stress hormone cortisol, as the immune system prepares for injury and infection.

Psychologists call these contrasting responses 'challenge' and 'threat'. When we face stressful situations in modern life – public speaking, a confrontation with someone we'd rather avoid, or a physical challenge such as a ski trail – the same ancient calculation comes into play. We subconsciously weigh our chances: deep down do we think we are going to win, or lose? The answer is generally a combination of factors, says Mendes. Have you studied for the test? Are you an optimistic person? Did you sleep well last night? 'All those factors can influence how we perceive our resources to cope with the task at hand.'

When it comes to longer-term health, challenge responses seem to be largely positive, while threat states are more damaging. Mendes has found that people who experience a challenge response bounce back to normal fairly quickly, and a range of studies suggests that mild to moderate bursts of such 'positive' stress, with time to relax in between, provides a useful workout for the cardiovascular and immune systems. 'In many ways, what we do in these psychologically stressful tasks parallels beautifully what you see in exercise stress,' says Mendes. Just as with physical exercise, if we put our bodies under a manageable amount of stress, then go home and rest, this eventually makes us stronger and more resilient. In essence this is what we're doing every time we go on a rollercoaster, or watch a scary film.

By contrast, people in a threat state take longer to recover to baseline once a task is over, both mentally and physically. They tend to worry more about how they did, and remain more vigilant for future threat. Their blood pressure also stays high. Over time, the extra strain on the heart can lead to hypertension. And as we've seen, repeated activation of cortisol can damage the immune system.

Intriguingly, Mendes has found that simply changing how we think about our physical response to stress can have a dramatic effect. She subjected volunteers to a gruelling ordeal called the Trier Social Stress Test. It involves 15 minutes of public speaking and mental arithmetic in front of a panel of stern judges, and in lab studies it reliably induces a state of fight-or-flight.

Mendes told some of the volunteers that experiencing physical symptoms of anxiety during the test, such as a racing heart, was a good sign. It meant that oxygenated blood was being delivered to their brain and muscles, she explained, and would help them to perform better. Remarkably, simply knowing this shifted these volunteers towards a challenge response – with greater vasodilation and cardiac output – compared to a placebo group (advised instead to ignore the source of their stress) and a group who received no instructions at all.[42]

In another study, Mendes showed that reframing the body's responses in this way doesn't just shift volunteers' physiology, it improves their performance too. She asked students preparing for the Graduate Record Exam (GRE) – a high-stakes test required for admission into graduate school – to sit a fake test in the lab. Compared to a control group, those told to interpret their stress as positive had the same physiological benefits as in the previous studies. But they also got higher scores – not just in the fake test but in the real GRE, which they sat up to three months later.[43] 'Of all my 60 or 70 published papers, this is the result that was most surprising for me,' says Mendes. 'It was such a tiny mindset change.'

Mendes' work shows that we don't have to be ruled by stress. Even with a small shift in attitude, we can start to reduce the health impact of stressful events, and to perform better under pressure. Unfortunately, it's not always easy just to decide to be less stressed or to think about our problems more positively. People who are chronically stressed, in particular, can become locked into negative patterns of thinking.

That's because, over time, stress physically rewires our brains.

***

One night at teatime, my five-year-old daughter leapt away from her fish fingers with a shout. She pointed to a sizeable spider on the wall by her chair, and refused to return until the creature was removed.

This posed a problem for me because I'm scared of spiders. But as the only adult in the house at the time it was up to me to do something about it. And (although I've clearly failed so far) I'm trying not to pass on my irrational fear to my daughter. I approached the offending arachnid, armed with a coaster and a plastic cup.

I could feel a fight going on in my mind. On one side was a red, flashing alarm. It contained no words, just deep-seated dread and revulsion. Doing battle with this primitive danger signal was a sensible, soothing voice that insisted everything was fine. These two armies were fighting for control of my body, too. One side urged my muscles to freeze, while the other was sending instructions to relax and advance. I duly ejected the spider from the kitchen, but doing it was like moving through treacle.

Most of the time, we maintain the illusion of being a coherent, whole person. But there are certain times, even during everyday experiences like confronting a spider, when the conflicting mechanics of the brain become exposed. When we sense a potential danger, several key brain regions interact to decide what we should do about it. One is the amygdala, a quick response system for detecting threats in the environment. It stores emotional memories, especially distressing ones, and when similar scenarios appear again, it triggers fear, anxiety and the fight-or-flight response. The source of phobias and prejudices, the amygdala works within a heartbeat and involves no conscious thought.

Working against its primitive drives are the hippocampus, which adds factual content to memories, and the prefrontal cortex, which carries out higher cognitive functions such as planning and rational thought. These work more slowly, but analyse situations more logically to defuse our alarm and turn down our response to fear or stress. Which side ultimately wins determines whether we lash out or speak kindly; whether we run or face our fears. And it turns out that in any particular person's brain, the odds are stacked in a way that depends on their life experiences, specifically their previous exposure to stress.

In one key experiment, psychologists showed short videos to teenagers at a high school in St Louis, Missouri. These featured neutral scenarios such as a sales assistant watching a shopper, and the students were asked to imagine themselves in each situation. Most saw nothing untoward, but those from disadvantaged backgrounds (after controlling for race) were much more likely to interpret the scenarios as threatening – thinking they were about to be accused of shoplifting, for example – and they experienced raised heart rates and blood pressure to match.[44]

This effect appears to last for life – Northwestern's Greg Miller got the same result when he showed the videos to adults raised in poor or well-off households.[45] Similar effects have been seen in caregivers like Lisa, and those who have suffered trauma or abuse in early life. People who are chronically stressed find small hassles much more stressful than normal. And they're much more likely to experience a threat rather than a challenge response.

Over the last few years, neuroscientists including Bruce McEwen of Rockefeller University in New York have discovered why. In animal experiments as well as in people who are chronically stressed, repeated activation of the amygdala causes it to become bigger and better connected over time, while the hippocampus and prefrontal cortex wither and shrink.[46] For example, a study carried out three years after the 9/11 terrorist attacks in New York found reduced grey matter volume in these brain regions in otherwise healthy adults living close to the destroyed buildings.[47] This reshaping of the brain has been linked to psychiatric disorders including dementia and depression.

Here, then, is one explanation for how the effects of early adversity can persist throughout a lifetime (we'll discover another in Chapter 10). Stress influences how the brain is wired, making us extra-susceptible to future problems by destroying the very brain pathways that would help us to stay calm and in control.

***

After meeting Susan, I head across town to a quiet road marked by a sign: 'Milledgeville Housing Authority'. The houses here are small bungalows, each divided into two tiny apartments. I'm struck by how impersonal they are compared to Susan's home. There are no borders, fences, flowers or garden furniture – just rows of identical brick boxes, evenly spaced on the grass.

I knock on the door of the address I've been given and am met by Monica. She takes a while to come to the door but greets me warmly. 'I forgot you were coming!' she says. The 39-year-old is wearing a strapless green-and-yellow sundress that leaves the expansive brown flesh of her chest, arms and shoulders to spill over the elastic. Her black hair is set into glossy curls, and when she smiles, a gold tooth flashes.

The front door leads directly into her living room, a small, square space with bare walls and a vinyl floor. The room is dimly lit – the blinds are closed despite gorgeous sunshine outside – and the only pieces of furniture are a faded blue sofa and chair, low table and TV. Despite the ashtray on the table, a few cigarette ends are scattered on the floor. Monica gestures to the sofa and absent-mindedly flips on the TV as we sit down to talk.

She tells me that she never graduated from high school and now works in a school cafeteria. She makes a face. 'I earn $700 a month,' she says. 'A month!' She's also single mum to Takisha, who is just home from school, dressed in red t-shirt and black leggings with a red bow in her long, plaited hair. The teenager is tall, but overweight and a little awkward. When instructed by her mother she sits opposite us, tapping buttons on her phone.

One of Monica's biggest worries is keeping her daughter safe, she tells me: 'I don't let her go anywhere.' Takisha is just 13, but already other kids in her class are smoking, drinking and having sex.

Monica recalls her own teenage years, including one evening when a close girlfriend invited her out. She didn't trust the other girl who was to join them, so she declined. 'The next day I heard they were locked up for robbery. They poured hot grease on this elderly guy and robbed him,' she says. 'Suppose I'd been in the car! One bad decision and that will change your whole life.' So far though, Takisha has stayed out of trouble and gets good grades in school (at one point in our conversation, she casually quotes in Latin), and she tells me she'd like to be a paediatrician when she grows up.

The pair clearly have a close relationship: they gently tease one another, and Takisha looks shyly towards her mother for approval before she speaks, for example when I ask how she spends her time. There isn't much to do in Milledgeville, it seems. 'I like being on my phone,' she says. 'I like eating.' Monica's answer is similar. The pleasures in her life are TV – she mostly watches talk shows, and documentaries featuring real-life stories, for example one about a teenager who hanged herself after being bullied online – and food. Takisha would eat healthy foods if she got the chance, says Monica, such as oatmeal, yoghurt or salad. 'But I don't eat it, so I don't buy it.'

Instead, she finds comfort in chicken wings and other fried foods. 'We are living in poverty,' she says. 'I turn to food. That is my everything. I hate it, but to drown out my issues and my stress I eat.'

***

Monica and Takisha aren't alone. In many different countries, scientists have found that people brought up in poverty are more likely to smoke and drink excessively and less likely to exercise. They eat unhealthy diets, and women are more likely to be obese.[48] As well as damaging health directly, these behaviours also make inflammation worse: smoking and a high-fat diet are both associated with higher inflammation, for example, while regular exercise can reduce it.

Why do people in poor communities behave differently? There are plenty of practical reasons: fresh vegetables and gym membership don't come cheap. There's also strong peer pressure to make bad life decisions; Monica has good cause to keep Takisha indoors even though that impacts her daughter's health. And for those with no realistic hope of ever escaping poverty to enjoy rewards such as a decent house, challenging job or fun holiday, or those who have regular experience of losing people or assets they value, perhaps focusing on cheap, immediate pleasures such as cigarettes or fried food is an entirely rational response.

But psychologists including Greg Miller think there is another factor, too. Research suggests that stress in early life doesn't just make people more vigilant for threat. It also affects reward circuits in the brain that regulate our appetites for everything from food to drugs, sex and money.

In addition to the amygdala, the prefrontal cortex helps to regulate other brain regions including the nucleus accumbens, which is part of an area called the ventral striatum. The nucleus accumbens makes us want things, and it plays an important role in addiction. Messages from the prefrontal cortex to the nucleus accumbens temper our desire, reminding us of the consequences of our actions, and helping us to forego immediate gratification for greater rewards in the future.

There's preliminary research suggesting early stress affects how these circuits too are wired as the brain matures, weakening this top-down control throughout people's lives. Those from low socioeconomic backgrounds are more likely to prefer smaller immediate rewards over larger postponed ones, regardless of their current life circumstances.[49] A 2011 brain imaging study asked 76 adults to play a game in which they could win or lose money.[50] When they learned of their winnings, those from poorer backgrounds had reduced prefrontal cortex activity, and weaker connections between the prefrontal cortex and the ventral striatum.

Someone with a brain wired in this way is likely to prioritise immediate pleasure over future consequences. They'll be impulsive, and at risk for unhealthy behaviours such as eating high-fat foods, addiction and risky sex. Like being hypersensitive to threat, this makes sense from an evolutionary point of view – if you're in an environment where resources are scarce and there are dangers everywhere, it's a good strategy to gorge on calorie-filled food when you find it, for example, or to breed young. But in the modern world, these behaviours make it harder for people to escape poverty, and at the same time, ruin their health.

In several different ways, then, stress can rewire the brain in a way that puts people struggling with adverse environments at an even greater disadvantage – and sets them up for a lifetime of chronic illness. This cruel legacy helps to explain why people who are exposed to stress, like Monica, make the choices they do, and why they still suffer health effects even if their circumstances improve. But it also raises a question. Can these changes in the brain be prevented, or even reversed?

9

ENJOY THE MOMENT
How to Change Your Brain

It's seven in the morning and I'm walking along the beach in Santa Monica, California. The low sun glints off the waves and the clouds are still golden from the dawn. Curlews and sandpipers cluster on the damp sand, while in the distance, white villas of wealthy Los Angeles residents dot the Hollywood hills.[1]

For half a mile or so, the beach is close to deserted. Then, just north of lifeguard station 27, I find what I'm looking for. In a neat line, a few metres back from the water's edge, a handful of people sit cross-legged on towels. They're members of a local Buddhist group, about to begin an hour-long silent meditation. I take my place at the end of the row, facing out to sea.

For centuries, followers of eastern religious traditions have meditated in search of spiritual enlightenment. The practice came to the west in the 1960s as part of hippie counterculture, endorsed by celebrities and bands such as The Beatles and The Doors. Since then it has grown in popularity as people seek peace and meaning amid the material concerns of modern life; it now seems no more notable to see meditators on a California beach than in a temple in Tibet.

I'm not here on a spiritual quest, however. I'm interested in scientific claims that meditating can improve physical and mental health by reducing stress. Of all areas of mind–body medicine, meditation, with its close links to religion and spirituality – not to mention mind-expanding drugs – has had a stormy relationship with science. Various studies since the 1970s have suggested that meditating monks can achieve a range of striking physical effects, from voluntarily reducing blood pressure to flooding their brains with highly synchronised electrical waves.

Some researchers, with close links to religious organisations, have been accused of finding what they want to see. And although monks who have spent much of their lives in remote retreats are undoubtedly capable of some stunning feats, it's not clear how relevant that is for the rest of us. In the last decade or so, however, a new generation of brain imaging studies and clinical trials has put meditation firmly on the scientific map. They're showing that although watching our thoughts might seem ephemeral, it can have hard physical effects on our brains and bodies.

But first, it's time to give this mysterious practice a try. There are hundreds of ways to meditate: compassion meditation involves extending feelings of love and kindness to fellow living beings (we'll learn more about this in Chapter 10); transcendental meditation has people focus on a repetitive mantra. Mindfulness, meanwhile, involves being aware of your own thoughts and surroundings. This is one of the most popular – and most studied – practices, so this morning I try a form of mindfulness meditation called open monitoring. Sit upright and still, and notice any thoughts that arise. Don't judge or react to them, just let them go.

I settle on my towel and start to contemplate the sparkling water. The view stretches out over thousands of miles of Pacific Ocean, and it is breathtakingly beautiful. Facing this vast expanse without thoughts and daydreams to fill my attention is slightly unnerving, however. My head is usually a tangle of ideas and words; spoken, written, heard, imagined, remembered. I'm not sure they will be so easy to banish.

***

I'm not alone in filling my head with abstract thoughts, says Mark Williams, emeritus professor of clinical psychology at the University of Oxford, UK. He's an expert in the psychological effects of meditation, and co-author of a 2011 book called *Mindfulness* that explains how training the mind to be more aware can reduce stress and anxiety in daily life. It became a surprise bestseller, with testimonials from celebrities such as Ruby Wax and Goldie Hawn.

'Most of us are preoccupied moment by moment, we're not actually aware of where we are or what we're doing,' he tells me. 'We're usually planning the future, or re-running something that has happened.'[2] When you're doing the dishes, for example, you might be thinking about the cup of tea you're going to have. When you're drinking the tea, you're planning your trip to the supermarket. When you're driving to the supermarket, you're thinking about what you're going to buy.

Instead of noticing our surroundings, we're caught up in our mental world. This can be a happy experience: daydreaming about a luxury holiday, perhaps, or planning the perfect birthday present for a friend. But we can also conjure up negative, stressful situations. We might be eating a delicious meal, bathing our kids or walking along a beach, but in our heads we're replaying yesterday's argument or stressing about tomorrow's work commitments, to an extent far beyond what's actually useful.

Getting lost in such brooding or worry in itself makes us stressed, but it also means that we fail to notice positive things in the world around us that might temper our anxiety. Getting ready for work in the morning, already immersed in the struggles of the day ahead, we're oblivious to the comforting warmth of our tea; a great song on the radio; our child's smile. 'You can live your life constantly missing your moments,' says Williams. We're in a bubble, cutting ourselves off from the small beauties and pleasures that make life worthwhile.

If we're not careful, mind and body can feed off each other in a downwards spiral, says Williams. Negative thoughts trigger stress responses in the body. But the process works in reverse too: when we're in fight-or-flight mode, the brain becomes hyper alert to threat. The more stressed we feel, the more likely we are to come up with negative thoughts.

Mindfulness meditation helps to stop that from happening. Becoming more aware of our own thoughts allows us to step back and realise that a negative or stressful notion doesn't necessarily represent reality, explains Williams. We don't have to respond emotionally. It's just spontaneous background chatter generated by the brain. And once we've recognised this, we can calm that chatter down.

Brain imaging studies support this idea. For example Giuseppe Pagnoni, a neuroscientist at the University of Modena and Reggio Emilia in Italy, scanned the brains of people experienced in Zen meditation, which, like mindfulness meditation, involves noticing thoughts and then dismissing them. Our internal monologue of spontaneous thoughts is thought to be generated by a set of brain regions called the 'default mode network', which is most active when we're not focused on any external tasks. Pagnoni found that the meditators could down-regulate the activity of this network, and that they were able to return to this calm state more quickly than inexperienced controls after being distracted.[3]

Having thoughts about the world has put us one step ahead of the zebra – but at a cost. We can become worn down by concerns over things that have already happened, haven't happened yet, or might never happen at all. Mindfulness, it seems, may put us another step ahead – we can have thoughts, but we don't have to be ruled by them.

***

At first, despite the gorgeous view, my mind seems desperate to be anywhere but on this beach. It's wriggling and darting, refusing to be silenced, throwing thoughts and images in front of me in quick succession. Eggs (I'm planning where to get breakfast), cab times (I've got a flight to catch), interview questions (I'm meeting a researcher this afternoon). Each one is calling me, tempting me to follow, to become lost in its twists and turns.

Every time I reject a chain of thought, another quickly follows, as if my mind is a market trader desperate to sell something: 'You don't like this one? Then try this!' A red jacket I bought the last time I walked down this beach. What presents to buy for my kids back home.

In an attempt to banish this mental whirlwind, I focus on the details of the scene in front of me, eyes fixed resolutely ahead. At first, the beach seems busy. Waves splash and splash, rumbling like gentle thunder. Sanderlings wheel along the shoreline. Joggers and dog walkers cross my field of vision, while groups of pelicans hang out on the water before taking wing or floating out of sight. A surfer, silhouetted black against the sky, bobs about for 20 minutes or so before he, too, is gone.

I'm immersed for a while but as time stretches on, I start to feel strangely detached from this shoreline activity. I imagine that the birds and joggers and surfers are like my thoughts; they inhabit different forms and timescales, but in the end, they all pass. They start to seem less important somehow, less real, and instead of watching them come and go, I find myself focusing further out towards the horizon. I'm drawn to its beguiling silence, and its unwavering line of deep, still blue.

By the end of the hour, my limbs are aching and the morning sun is hot on my cheek. As I look up and down the beach after this first foray, I do feel calm and strangely connected; more a part of the larger landscape, perhaps, and less concerned with the personal minutiae of my day. I like the idea of being free of negative thoughts (who wouldn't?) and I can see that over time, this might be a potent technique for gaining a different perspective on life. But does it really work? Most of us aren't monks, and we can't meditate all the time. Can a few short sessions really protect us from – or even reverse – the ravages of stress? And can that in turn affect our physical health?

***

Gareth Walker knows something about how the past and future can torment us. Ten years ago, he was working in Sheffield, northern England, as a police officer, or as he puts it, a 'bobby on the beat'. The 26-year-old enjoyed his active job, and in his spare time he loved going out into the hills and walking through the beautiful Yorkshire dales.

Then one morning in 2006, Gareth woke to find that the vision in his left eye was blurry. His optician couldn't find anything wrong. His doctor prescribed antibiotics for conjunctivitis, but it made no difference. Eventually he had MRI scan, and the neurologist dropped a bombshell: Gareth's immune system was attacking his optic nerve. The most likely explanation was that he was suffering from multiple sclerosis (MS).

A chronic condition in which aberrant inflammation gradually destroys the nervous system, MS can cause wide-ranging symptoms as a patient gradually loses control over his or her body. One by one, limbs, eyes, bowel and bladder can stop working. Patients also suffer pain and fatigue, as well as cognitive and emotional problems – especially depression. MS generally starts as a 'relapsing remitting' form, in which attacks come on suddenly then fade. Eventually, however, the condition becomes 'progressive', meaning that the damage steadily gets worse and worse. There are few effective treatments and no cure.

It takes two such bouts of inflammation to diagnose MS, because occasionally people have just one attack then suffer no further problems. But if anything else happened, Gareth's neurologist warned, he would be faced with inexorable, worsening disability. For three years, Gareth tried to continue living a normal life. Then, in 2009, he started to lose control of his bladder. In 2010, he was formally diagnosed with MS.

He describes the time following the diagnosis as 'horrifically stressful'. Soon afterwards, Gareth started to have trouble walking, and had to take sick leave from the job he loved. And that June, he became a father.

He recalls that in August 2010, he took his wife and son for a week's holiday, staying in a cottage in the picturesque village of Tosside. It was a chance to get away as a family and to celebrate the new baby. One sunny day, they had a picnic in a nearby nature reserve, eating their sandwiches on a bench by a stream. Gareth's wife suggested they walk down to the water, just a few feet away down a shallow but slightly rocky path. As Gareth began to pick his way across the rocks, he felt unsteady on his feet, as if he might fall.

Suddenly it hit him. If he was struggling now, how would he be in a few years' time? He looked at his precious two-month-old boy and was struck by the thought that he would never join his son on walks like this. He would never skim stones with him, or play football with him, or countless other things that normal dads do with their kids. Instead, he'd be in a wheelchair, crippled and helpless. That single moment ruined what could have been an idyllic family celebration, and started an avalanche of fears and imaginings that he couldn't escape.

'Every dream that I'd had about the future was suddenly taken from me,' he says. 'I didn't know what to do. It was a very, very tragic time for me.'[4]

Five years on, however, Gareth is apparently far from despair; in fact he says he is happier than he has ever been. He credits mindfulness meditation with transforming his life and is now one of its most influential advocates, with his own website and more than 60,000 followers on Twitter. I've arranged to meet him to find out more about this incredible turnaround.

He picks me up from the train station in his home of Barnsley, a former mining town in the heart of Yorkshire. It's lunchtime on a cold January day, and he immediately drives me out of town past snow-coated fields, to the rural village of Silkstone. There's nowhere nice to eat in Barnsley, he apologises.

Now 36, Gareth is friendly, relaxed and down to earth, with steady grey eyes and flat northern vowels. Dressed in a red jumper and jeans, he's slim but not to the point of frailness. He chats – about mindfulness, he could talk about its benefits all day, he says – until we arrive at our lunch spot. He looks for a disabled space, then uses a crutch to help him across the short distance from the car park to the café.

When we're settled, I ask Gareth how he discovered mindfulness. Shortly after that disastrous walk to the stream, he says, someone recommended it to him as a way to cope with the stress of being diagnosed with MS. 'I didn't have a clue how to meditate,' he tells me. 'I'd heard the word but I just thought it was something for hippies.' So he picked up a bestselling guide to mindfulness by a US author named Jon Kabat-Zinn: *Wherever You Go, There You Are.*

He started by meditating five minutes at a time, in a fairly ad hoc way. He would close his eyes and count his breaths in and out. If he had a thought before he got to ten, he went back to the beginning and started again. Nothing much happened at first. But after a few months, he noticed a change.

***

If Elizabeth Blackburn leapt across a canyon between psychiatry and biochemistry in her study of telomeres, the rift that Kabat-Zinn tackled was even wider. A molecular biologist and yoga teacher from Massachusetts, he was convinced that the meditation he practised as part of his Buddhist faith could help people for whom there was little that physicians could do; those who were dying, for example, or ravaged by pain. But he knew that doctors would never prescribe a religious practice. Then one day, while meditating on a retreat, he had a vision. He would reinvent mindfulness, stripping out its religious and spiritual aspects, to make it palatable to the medical profession.

In 1979, he developed an eight-week course that included elements of mindfulness meditation as well as relaxation techniques and hatha yoga. He called it mindfulness-based stress reduction (MBSR), and founded a clinic at the University of Massachusetts in Amherst. 'He told the doctors in the hospital, give me your patients for whom you have no hope,' says Trudy Goodman, who runs InsightLA, the Buddhist group I joined on Santa Monica beach, and who worked with Kabat-Zinn at the time. 'People sent them, they didn't know what else to do with them. Some people's pain eased. Some people died peacefully.'[5]

Secularising meditation was a risky strategy at the time. 'People said, "You are watering down the teachings, it will come to no good,"' says Goodman. 'It was unheard of to extract mindfulness from the Buddhism that we were all studying.' But it transformed the practice from a niche religious method into a cultural phenomenon.

Since Kabat-Zinn founded his clinic, more than 20,000 people have completed his eight-week programme. MBSR has been featured in countless newspaper and magazine articles, and in television programmes including the Oprah Winfrey Show. According to the National Institutes of Health (NIH), nearly one in ten American adults now meditate.[6] There is a dedicated monthly magazine, called *Mindful*, and hundreds of mindfulness apps. Searching for 'mindfulness' on Amazon turns up nearly 19,000 books and DVDs, from spiritual journeys to practical stress reduction plans and even exercises for kids, while mindfulness sessions and lectures are run everywhere from Silicon Valley to Capitol Hill.[7]

This is in large part because distancing mindfulness from its religious origins has opened the door to scientific studies of its potential benefits, further legitimising the technique. There have now been hundreds of randomised controlled trials of mindfulness-based therapies. Systematic reviews and meta-analyses consistently conclude that MBSR can reduce chronic pain and anxiety, and that it reduces stress and improves quality of life in everyone from cancer survivors to healthy volunteers.[8]

Some are concerned by this explosion in popularity. Some Buddhist teachers complain that mindfulness has become commercialised, and that the subtleties of what it should mean have been lost.[9] Psychologists have warned that mindfulness classes are increasingly being offered by under-qualified teachers presenting themselves as experts, while news headlines describe tragic consequences in vulnerable participants attending meditation retreats: during one such course in Arizona desert, for example, participants had to meditate for long periods of time without food or water before attending a ceremony in a 'sweat lodge' – three died and 18 were hospitalised with injuries ranging from heat exhaustion to kidney failure.[10]

Meanwhile Kristin Barker, a sociologist at the University of New Mexico in Albuquerque, sees the movement as one enormous guilt trip, describing mindfulness meditation as 'do-it-yourself medicalization of every moment'.[11] She highlights advice from Kabat-Zinn such as to meditate 'as if your life depended on it, because it does to such a profound extent'. The idea that our health depends on being mindful all the time turns all of us into patients who need treatment to fix our unhealthy thoughts, she warns, and makes us feel like failures if we don't achieve this blissful state.

Gareth laughs off that last point. 'Nobody can be mindful all the time,' he says.[12] After a few months of meditating for just five minutes a day, though, he says he started finding it easier to stay in the present moment. As a result he felt more patient, and was less frustrated by physical challenges such as climbing the stairs. 'If I could not get too far ahead of myself and just stay in the moment, things got a lot easier,' he says. After that, he started meditating for longer periods, and claims that the benefits were 'astronomical'.

Most of the agony in having MS comes from the past or the future, he explains. After being diagnosed, he was tormented by thoughts about all the things he used to love – his job, rambling – that he would never do again, and about the future, such as the fear that if MS takes his eyesight, he won't see his kids grow up (he now has two sons); or that he will have to suffer unbearable pain.

'I have to pull myself back from those thoughts countless times in any given day,' he says. And he believes that his regular mindfulness training makes this easier to do. 'I'm only a 36-year-old man, how on earth am I going to be in ten years' time? That story starts but I never let it get any further.' If he can stay in the present moment and focus on what's happening around him, he says, most of the agony of his condition goes away and life is good – even great.

Gareth now meditates for half an hour every day. He sets his alarm early and meditates sitting up in bed, either focusing on his breath, or with headphones on, keeping his attention on the music. But he also tries to integrate mindfulness into his life. 'If my son comes up and interrupts then he becomes the subject of the meditation.' This means that instead of allowing himself to become distracted while playing, he gives all of his attention to his son.

As well as helping him to enjoy and appreciate the life he has, including time with his children, Gareth believes that mindfulness has helped him to become more tolerant and empathic: 'You can only empathise with someone when you notice things – like a frown on your partner's face – and mindfulness is about noticing things.'

It also helps him to cope with pain. Gareth suffers from trigeminal neuralgia: episodes of intense stabbing pain – like an ice pick, he says – into the side of his face. These are predicted to get worse as his condition progresses. He tells me a Buddhist story about how pain has two arrows: the physical pain, and then the story we attach to that pain. The metaphor reminds me of the burns patients we met earlier whose pain is magnified by anxiety and fear. But rather than distracting themselves from pain with a tool like Snow World, mindfulness meditators aim to remove the emotional content by meeting pain head on.

'You let the pain in,' explains Gareth. 'You embrace the pain, you invite it in for a cup of tea and give it a cuddle. It sounds crazy, but it really works. The effects of the episodes of pain that I have are far, far less.'

What about fatigue, I ask? It's usually a big problem for MS patients. Gareth says he used to feel exhausted but not since he started meditating. He now leads a busy life by any standards; as well as juggling parenthood and coping with his condition he is back at work full time – now at a desk job investigating complaints against the police. And he still runs his website, Everyday Mindfulness,[13] with its associated Twitter account. (His most retweeted quote is from the Buddha: 'Pain is inevitable; suffering is optional'.)

'People think of meditation as a time consumer, but the opposite is true,' he says. 'It is a time provider, because of all the time that we don't spend following useless trains of thought. I wouldn't be able to lead the life that I lead now if it weren't for meditation.'

I'm not sure I really 'got' the point of meditation until talking to Gareth. It's not a quick fix; it requires hours of regular practise, and more trials are needed to establish exactly who it helps and how. But here, munching sandwiches in the snowy Yorkshire dales and listening to this father and police officer describing daily pain, stress and fear that dwarfs anything I have to deal with, I can't help thinking that if mindfulness allows him to face his demons with courage, and even joy, then it must be a pretty powerful tool.

***

It's a bright February morning, and I'm jiggling about self-consciously in a room full of strangers. This is the Mood Disorders Centre at the University of Exeter, UK, and those here with me hope that by heading off stressful thoughts, mindfulness can protect them against the life-threatening despair of major depression.

The course is called Mindfulness-Based Cognitive Therapy (MBCT). Developed by Oxford University's Mark Williams and his colleagues, it is largely based on MBSR but with a focus on depression. The conventional medical view is that depression results from a chemical imbalance in the brain – a lack of the neurotransmitter serotonin. Most antidepressants boost serotonin levels. But drugs alone help only around a third of patients out of depression, and as we saw in Chapter 1, much of their benefit is actually down to the placebo effect. And, like most drugs, they are associated with side effects (from gut problems and sexual dysfunction to suicidal thoughts).

Psychological therapies are an increasingly popular alternative. The best studied is cognitive behavioural therapy (CBT), in which therapists talk to patients about their lives and problems, aiming to help them identify negative, unhelpful thought patterns and to change them. But MBCT (which combines mindfulness with some elements from CBT) is fast catching up. Whereas CBT is an acute treatment for those who are already ill, MBCT is designed as a tool that people can use in everyday life, to help them stay well. Today's session is a refresher for people who have finished the course, run by psychologists Willem Kuyken and Alison Evans.

There are 30 people here of varying ages and backgrounds, and they have all suffered from recurrent bouts of severe depression in the past. Evans takes us through some different exercises, punctuating each by striking a resonant metal bowl. After focusing on our breathing, she asks us to attend to our bodies and any physical sensations we're feeling. Then comes the more active jiggling about. The idea is that paying attention to your body's movements helps you focus on the present, rather than getting caught up in worries about the past or future.

'You're seeking safety in the moment,' explains Kuyken[14] from the front of the seminar room – tall, chiselled, intense. 'And if you can cope with this moment, that shapes the next moment.' Trainees are encouraged to carry this principle into their everyday lives – going for a walk and paying attention to the trees and sky, for example, or just breathing – to escape negative patterns of thinking that threaten to overwhelm them. Another trick is to use everyday cues – red traffic lights, opening the fridge – as a reminder to be aware and notice their surroundings.

Trial results for MBCT so far are impressive. In studies published in 2000 and 2004, Williams and his colleagues found that MBCT cut the relapse rate in patients with recurrent major depression in half.[15] This caused the therapy to be recommended by the UK's National Institute for Health and Care Excellence (NICE). Then in 2008 Kuyken carried out a further study, reporting that patients given MBCT had fewer symptoms, better quality of life and a lower relapse rate than those on drugs.[16]

The patients here in Exeter seem convinced by the benefits of mindfulness. 'I hated antidepressants,' says Vicky, 43, a short, practical woman who has suffered from depression for 20 years. 'I'd always get off them as soon as I could, then move on and try to forget that time. But a little thing would provide a trigger and I'd be down in the depths again.' Each time it happened, she says, it was worse than before, and harder to hide from her kids. When she was depressed, she wouldn't want to get out of bed for days on end.

Vicky finished the course two years ago, and says it has helped her to notice warning signs – things like being constantly busy, not sleeping properly, feeling anxious all the time – that indicate she could be on the brink of relapse. Before, 'I couldn't understand why I had suddenly fallen down a mineshaft,' she says. 'Now I'm much more mindful of how I'm feeling. It's like a safety ladder that can get me back out of that shaft.'

Another attendee, 33-year-old Sue, is a keen rock climber and had a promising career as an oceanographer until being bullied at work triggered a bout of severe depression. 'It was like a switch flicked,' she says. 'I'd get so worked up my heart would be racing. I'd be sweating, nauseous, I couldn't get out of the front door.' After being prescribed antidepressants for an episode of depression ten years earlier, Sue vowed never to take the drugs again. 'They're really hard to come off, and have horrible side effects. And they don't solve the underlying problem.'

She had a course of CBT, before being referred to Kuyken. With CBT, 'You try to stop having "wonky" thoughts,' she says. 'But it's easy to beat yourself up, to think that you've been feeling the wrong way.' With mindfulness, 'I was relieved not to be talking through everything,' she says. 'This is more accepting. It's not your fault.' She does have some concerns – for example that the constant sparking of ideas in her head might have been necessary for her creative ability as a scientist. But she approaches mindfulness as just another experiment. 'If I can't do something, I'll do three minutes of breathing and try again. It's amazing what a difference it makes.'

Then there's Ann, a 57-year-old with a wrinkled face and white hair in a ponytail, who has suffered from recurrent depression for most of her life. At her lowest she was suicidal, believing that her children would be better off without her. She too hated being on antidepressants. 'They zombified me,' she says. 'They didn't just cut off the negative feelings, they cut off all feelings.' Now she meditates every day and is confident that it will help her to stay well without the drugs. 'I've realised that thoughts cannot harm you.'

When I ask how her life is different because of MBCT, her reply is simple: 'I'm alive.'

Once the session is over, Kuyken and I sit in his sunlit office, where he tells me he's hopeful that MBCT can be adapted to help people suffering from other mental conditions too, such as chronic anxiety, social phobia or eating disorders. Ultimately, however, he believes that mindfulness might help all of us to cope with the demands of modern society.

'We're increasingly living our lives on autopilot,' he says. 'Kids are getting mental disorders younger and younger.' In particular, he believes that always-on technologies such as email, mobile phones and Facebook can be harmful if we don't learn how to control their effects on us. 'We're constantly having to deal with incoming data.' It's very hard to do that with awareness, he says, responding thoughtfully to what's happening around us rather than reacting blindly.

First, however, he hopes to amass even stronger evidence for the benefits of MBCT in patients with recurrent depression. At the time of writing, he and his colleagues have just published a trial that followed more than 400 patients for two years: MBCT protected them against relapse just as well as antidepressant drugs.[17] (When pooled with data from previous trials, patients who took a course of MBCT were 24% less likely to relapse than those on medication.)

'There are millions of people with depression in the world,' says Kuyken (who has moved since I visited Exeter and is now director of the Oxford Mindfulness Centre). 'If we can provide them with an alternative to antidepressants, that's huge.'

It's a long way from where he started, back in 2000, when he nervously 'came out of the closet' to study MBCT after a long-standing private interest in meditation. Williams, too, says he initially feared that even to study meditation might destroy his academic reputation: 'When we did the first trial we thought we would meet enormous scepticism. Part of me was worried that my career might suffer. But scientists were really interested.'

That positive attitude is in large part because of a spate of recent findings that is now forcing scientists to take it seriously as a phenomenon with impressive physical effects. So I travel to Boston, Massachusetts, to meet the woman who has perhaps done more than anyone else to show what meditation does to the brain.

***

'I used to think the whole mind–body stuff was nonsense. But after a month of yoga class, I was hooked.'

Harvard neuroscientist Sara Lazar has bare feet and is sitting cross-legged on her chair. Her unruly hair is grey, but she has the energy and enthusiasm of a teenager. She laughs a lot and talks so fast that she skips some words altogether. 'It rocked me. I could see there was way more to it than just stretching and exercise.'[18]

We meet in Lazar's office in Boston's Navy Yard, unremarkable except for a shelf above her desk, which is adorned with pink blossoms in a tall, green vase; a bronze Buddha figure; and a silver dancer in a seated yoga pose – leaning forward with one leg straight and one bent. 'I like that one,' she says. 'I was doing that pose in yoga when I had this big "Aha!" moment.' Instead of struggling to push herself into the position as usual, she relaxed. 'I went like three inches further,' she laughs. 'Relaxing gets you further than stressing and straining!'

At graduate school, Lazar studied bacterial genetics. Then she injured her knee while training for a marathon. Temporarily unable to run, she started doing yoga to keep fit, and was amazed by the effect it had on her. Like Gareth, she felt that her brain was working differently. 'It changed how I thought about things,' she says. She felt that she was calmer, had more empathy, and was better able to see different points of view. 'I live in Boston, where there are lots of crazy drivers,' she says. 'I realised that I don't have to get angry with them. They're probably in a hurry, they're probably stressed too.'

Fascinated by what was happening to her brain, Lazar switched from bacteria to neuroscience. She trained in the use of magnetic resonance imaging (MRI) – the same technology that I watched doctors across town at the Boston Medical Center use to image the cyst in Daniel's brain. It isn't possible to do yoga in the cramped confines of a brain scanner, so instead she investigated the related practice of meditation.

She describes her decision to enter the world of mind–body medicine as 'brave or crazy'. 'Everyone sort of looked at me funny,' she says. At the time, in the late 1990s, meditation was seen as a hippie, drug-related practice, not a suitable subject for scientific research. But at around the same time, the NIH created a national centre for alternative and complementary medicine (the same one that got placebo researcher Ted Kaptchuk hired at Harvard). 'That gave me the confidence that I could do this, and I would get funding.'[19]

Other researchers were already studying how meditation affects brain activity, notably neuroscientist Richard Davidson at the University of Wisconsin–Madison. The Dalai Lama sent Davidson eight of his most experienced monks, each of whom had meditated for tens of thousands of hours.[20] Compared to student volunteers, when the monks meditated Davidson saw a dramatic increase in high-frequency brain activity called gamma waves – higher than neuroscientists had ever reported before (in healthy brains anyway; very high gamma waves are also seen during epileptic seizures).

This surge of gamma waves suggested that when the monks meditated, their brains were very highly organised and co-ordinated, with large numbers of neurons firing together. They also had very high activity in their left prefrontal cortex, a region linked to positive thoughts and emotions. The results were intriguing. These seasoned meditators were clearly able to induce states of consciousness outside of the normal realm of experience.

Lazar did something different, however. She was convinced that her yoga practice had induced not just a passing state of consciousness but a permanent shift in how her brain worked. 'I knew that my brain had changed,' she says. So rather than looking at the activity of the brain, she probed its physical structure. She didn't have access to monks, so she studied 'average joes' from Boston – a therapist, a chef, a lawyer, an IT guy – who were experienced meditators and practised daily.

To show me what she found, Lazar pulls up a series of scans on her computer screen. She must have seen tens of thousands of these over her career, but she's still wide-eyed with wonder at this window inside the human skull. 'It astounds me that you can get this detailed picture of a brain,' she says. 'There are some that look crystal clear, it's amazing.'

Lazar is astounded by what we can see in these images, but I'm also struck by what we can't. This is a human being, yet in these intricate, detailed structures there's nothing to tell us who this person cares about, their first memory, the music they love, the food they hate. We still have an achingly long way to go in our understanding of the brain. For now though, these black-and-white images are the best window we have on its secrets. What mark would meditation leave?

Lazar published her results in 2005. Compared to a control group, the meditators' cerebral cortex, including the prefrontal cortex, had grown in thickness by around a tenth of a millimetre.[21] 'The change is really tiny,' says Lazar. 'But it's significant.' It was enough to show that rather than just being a transient state, meditation can change the physical structure of the brain.

'That really shook everything up,' says Lazar. Scientists had only just worked out that it's even possible for the adult brain to change in response to its environment. It was long thought that by the time we reach adulthood, our brains are on a downward trajectory. Neurons can die, but they can't be born. In 1998, however, post-mortems on the brains of elderly cancer patients showed that new cells were being created even at the end of their lives.[22]

After that, studies started to show that everyone from violinists to taxi drivers beef up relevant brain areas with new cells and connections, just as we build muscles with physical exercise. Lazar's study showed that meditation can do this too. For the first time, it was possible to explain how the practice might permanently change psychology and physiology.

Other researchers followed, reporting similar results for several different kinds of meditation. There was still a problem, however. These studies left open the possibility that 'meditators are weird', as Lazar puts it. Perhaps people who choose to meditate have particular lifestyles (many of them are also vegetarian, for example) that may affect their brains, or perhaps people with certain types of brains are more likely to meditate in the first place. To prove that meditation was causing the changes, it would be necessary to take people who had never meditated before, and see how the practice affected them.

So that's what Lazar did, in two studies published in 2010 and 2011. Compared to a control group, those who took an eight-week course of MBSR had increased grey matter in brain areas involved in learning, memory and emotion regulation, including the hippocampus. They also felt less stressed, and this change was accompanied by reduced density of grey matter in the amygdala.[23]

'That's important,' says Lazar. As we saw in Chapter 8, chronic stress and depression leave people with a smaller hippocampus and prefrontal cortex and a bigger, better-connected amygdala. After just eight weeks' training, Lazar was seeing some of those changes in reverse. Her finding suggests that meditation can stack the odds back in our favour, making us more resistant to stress.

Lazar is now running a study to test whether physical exercise (which also reduces stress) causes similar changes. And she's investigating the potential of meditation for warding off dementia. The hippocampus and prefrontal cortex tend to shrink as we age, mirroring some of the changes caused by chronic stress, and contributing to cognitive decline. Several studies have hinted that meditation might help to slow this process. The difference in cortical thickness that Lazar saw was most marked in older meditators, while several different teams have now found that cognitive performance and grey matter volume decline more slowly with age in meditators than they do in controls.[24]

In a study published in 2014, Lazar also found that fluid intelligence (a measure similar to IQ) declines more slowly with age in yoga practitioners and meditators compared to controls, and that different areas of their brains remain more connected.[25] 'That's part of fluid intelligence and that's what goes with age,' she says. 'It suggests that meditation is helping to keep those regions talking to each other.'

Lazar's research is part of a huge NIH effort to find ways to prevent and treat Alzheimer's as the population ages. Her decision to study meditation might have seemed crazy at the time, but she's part of the establishment now.

***

I'm now persuaded that in those who practise regularly, at least, mindfulness meditation has the potential to change both our minds and our brains. But I still want to know – do those stress-busting effects reach beyond the brain to influence our immune system? And could being mindful possibly help to slow the progression of an autoimmune disorder like MS?

Back in the Yorkshire café, I ask Gareth what he thinks. He tells me that in 2011, shortly after he began meditating, he was diagnosed with the more serious 'progressive' form of MS, in which instead of having periodic attacks, people just get steadily worse. But in the five years or so since then he has surprised his doctors, because his condition has remained largely stable.

When he suggests to them that his meditation practice might be helping to slow the progression of his disease, Gareth says he gets 'a scornful look'. But he's convinced that mindfulness is a factor: 'I've had progressive MS for five years, I should be worse than this.'

There is growing acknowledgement, however, that by causing chronic inflammation, stress does exacerbate the progression of autoimmune conditions like MS. A 2004 meta-analysis of 14 studies published in the *British Medical Journal* concluded that there is a 'consistent' and 'clinically meaningful' association between stressful life events and subsequent attacks in relapsing remitting MS.[26] For example, a Dutch study that followed 73 MS patients found that stressful life events – such as redundancy, or the death of a relative or friend – doubled the risk of exacerbation over the following month.[27]

And in 2012, a randomised controlled trial of a stress management therapy in 121 patients with relapsing remitting MS found that those in the stress management group had fewer new brain lesions (a sensitive marker of the disease's progression) than the controls.[28] The effect size was similar to that seen in equivalent trials of new drugs. But the benefit didn't last once the therapy finished – six months later, there were no longer any differences between the two groups.

By giving people something to practise long-term, might mindfulness have more lasting effects? There are now lots of studies suggesting that mindfulness meditation does reduce signs of physiological stress in the body, such as the hormone cortisol, and markers of inflammation. Meanwhile some small studies, including a three-month meditation retreat studied by Elissa Epel and Elizabeth Blackburn, hint that meditation can protect or even lengthen telomeres, potentially slowing ageing in our cells.[29]

That's a dramatic finding, but not everyone is convinced. David Gorski, an oncologist at Wayne State University and critic of alternative medicine, warns that early results regarding the benefits of meditation are being oversold, particularly because with meditation, as with other mind–body therapies, it is impossible to carry out double-blind trials. 'Are you doing it rigorously?' he says. 'It is easy to be led astray. Nobel Prize-winners are not infallible.'[30]

Some scientists are still 'very uncomfortable' with the idea of studying meditation, responds Blackburn. She says she always emphasises that her studies so far are preliminary, but people 'see the newspaper headings and panic'.[31] To persuade the sceptics, she'll have to show the effects in larger studies. She and Epel are now working on a two-year trial of more than 180 mothers of children with autism (Lisa is one of them), to see whether a mindfulness course helps to protect their telomeres against the effects of stress.

Other evidence of meditation's effects on physical health is mixed. Kabat-Zinn reported in 1998 that the autoimmune skin disease psoriasis cleared more quickly when conventional treatments were combined with MBSR.[32] Other trials have suggested that MBSR boosts the response to flu vaccine,[33] for example, and reduces the number of colds that people suffer over the winter.[34] But most of these findings still need to be replicated before they can be believed.

Very few studies have looked at mindfulness for MS. A 2014 meta-analysis found just three trials; they showed significant benefits for quality of life and mental health as well as depression, anxiety and fatigue.[35] None has yet looked directly at disease progression, but the author of that meta-analysis, Robert Simpson of the Institute of Health and Wellbeing at the University of Glasgow, says it's something he'd love to look at in the future.[36]

Whether or not mindfulness turns out to influence the physical progression of his disease, however, Gareth says the psychological benefits alone make it worthwhile. In fact, despite suffering from a condition in which a very high proportion of patients become clinically depressed, Gareth insists he's happier now than at any time in his life. 'My wellbeing is wonderful,' he tells me over our coffee. 'MS makes some things very, very difficult. But life is difficult. I prefer to focus on the good things, and I have so many of them.'

He recalls the day he tried to walk down to that stream with his baby son: when his fear of the future sent him into a spiral of despair; when the happiness of an entire day was wiped out by just one thought. 'Now if that happened, I'd say, "Okay, it's just a thought,"' he says. 'And I'd struggle down to the stream and enjoy that moment.'

10

FOUNTAIN OF YOUTH
The Secret Power of Friends

The Nicoya peninsula in north-western Costa Rica is one of the most beautiful places on the planet. This 75-mile sliver of land, just south of the Nicaraguan border, is covered with cattle pastures and tropical rainforests that stretch down to the crashing waves of the Pacific Ocean. The coastline is dotted with enclaves of expats who fill their time surfing, learning yoga and meditating on the beach.

For the locals, life is not so idyllic. They live in small, rural villages with limited access to basics such as electricity, linked by rough tracks that are dusty in the dry season and often impassable when it rains. The men earn a living by fishing and farming, or work as labourers or *sabaneros* (cowboys on huge cattle ranches), while the women cook on wood-burning stoves. Yet Nicoyans have a surprising claim to fame that is attracting the attention of scientists from around the world.

Their secret was uncovered in 2005 by Luis Rosero-Bixby, a demographer at the University of Costa Rica in San José. He used electoral records to work out how long Costa Ricans were living, and found that their life expectancy is surprisingly high.[1] In general, people live longest in the world's richest countries, where they have the most comfortable lives, the best healthcare and the lowest risk of infection. But that wasn't the case here.

Costa Rica's per capita income is only about a fifth that of the United States, but if its residents survive the country's relatively high rates of infections and accidents early in life, it turns out that they are exceedingly long-lived – an effect that is strongest in men. Costa Rican men aged 60 can expect to live another 22 years, Rosero-Bixby found, slightly higher than in western Europe and the US. If they reach 90, they can expect to live another 4.4 years, six months longer than any other country in the world.

The effect is even stronger in the Nicoya peninsula,[2] where 60-year-old men have a life expectancy of 24.3 years – two-to-three years longer than even the famously long-lived Japanese. Nicoya is one of the country's poorest regions, so their secret can't be better education or healthcare. There must be something else.

Another longevity expert, Michel Poulain of the Estonian Institute for Population Studies in Tallinn, travelled to Nicoya with the journalist Dan Buettner in 2006 and 2007 to investigate Rosero-Bixby's findings.[3] The pair were working for the National Geographic Society, identifying long-lived communities around the world – which they dubbed 'Blue Zones' – and attempting to work out their secrets. Other examples included Sardinia, Italy, and Okinawa, Japan.

In Nicoya, Poulain and Buettner met people like Rafael Ángel Leon Leon, a 100-year-old still harvesting his own corn and beans and keeping livestock, with a wife 40 years his junior. Living nearby was 99-year-old Francesca Castillo, who cut her own wood and twice a week walked a mile into town. And there was 102-year-old Ofelia Gómez Gómez, who lived with her daughter, son-in-law and two grandchildren. When Buettner's team visited, she recited from memory a six-minute poem by Pablo Neruda. All of the elderly people they saw were still mentally, physically and socially active, despite their advanced age.

Poulain and Buettner drew up a list of things that might be helping Nicoyans to age so well. They have active lifestyles, even in old age. They have strong religious faith. The lack of electricity for lighting means they go to bed early, sleeping an average of eight hours a night. They drink calcium-rich water (which is good for the heart) and eat antioxidant-rich fruits.

Although the project was intriguing, it couldn't narrow down the crucial factors. But Rosero-Bixby has recently carried out a study aiming to do just that. He teamed up with David Rehkopf, an epidemiologist at Stanford University in California. The pair took blood samples from around 600 elderly Costa Ricans, including more than 200 from Nicoya. They sent the blood samples to Elizabeth Blackburn's lab in San Francisco, where she measured the length of their telomeres. If the Nicoyans really were ageing more slowly, it should show up in her results.

The team reported in 2013 that Nicoyans' telomeres are indeed longer than those of other Costa Ricans.[4] Their impressive life expectancy isn't a statistical fluke but a real biological effect, in which their cells look younger than expected for their age. The size of the effect was equivalent to changes caused by behavioural factors such as physical exercise or smoking.

To investigate why the Nicoyans' telomeres are so long, Rosero-Bixby and Rehkopf analysed the effects of everything from the residents' physical health and level of education to their consumption of fish oils. Diet makes no apparent difference, and the Nicoyans are worse off than other Costa Ricans when it comes to health measures such as obesity and blood pressure. Their slower ageing doesn't seem to be a consequence of genes either – Nicoyans lose their longevity advantage if they move from the region. And it isn't money: richer individuals actually have shorter telomeres.

But there are some clues. Rehkopf and Rosero-Bixby found that Nicoyans are less likely than other Costa Ricans to live alone, and more likely to have weekly contact with a child. Such social connection seems crucial. The telomere length difference is halved among Nicoyans who don't see a child each week, and if they live alone, they lose their advantage completely.

Other studies have found that Nicoyans have greater psychological attachment to family than residents of Costa Rica's capital, San José. So Rehkopf and Rosero-Bixby speculate that close family ties might protect Nicoyans against life stress that would otherwise shorten telomeres. Despite their poverty, strong social bonds keep them young.

It's a startling finding, and to confirm it will take studies that collect more detailed data about the Nicoyans' social connections. But Poulain says the theory fits with his own observations. He emphasises (as does Rehkopf) that there is no single secret to long life, and that residents of longevity hotspots such as Nicoya probably enjoy a lucky combination of genetic and environmental factors. Yet he has seen unusually strong social networks in other Blue Zones too. 'The social aspects are crucial,' he says. 'There's terrific support for the elderly.'[5]

The idea is also bolstered by decades of evidence from communities suffering the reverse phenomenon: the gradual loss of social ties.

***

The south London council estate where 69-year-old Lupita Quereda lives is grim and grey, all paving slabs and concrete. I'm visiting her with a staff member from the charity Age UK, which sends volunteers to chat with isolated elderly people. The shared stairwell up to Lupita's home is covered in dirt and cobwebs, and there are multiple locks on her door.

But she opens with a huge smile, welcomes us in, and ushers us to a plain wooden table in her tiny kitchen. The flat is neat and tidy, with walls painted warm, orangey-red. There's an old-fashioned cooker, and the kitchen shelves hold piles of audio tapes, a few pumpkins and squashes, and a South American wooden doll. Lupita wears a nightdress (she finds loose clothes more comfortable after a recent fall) and a maroon housecoat. She has graceful hands and elegant features but her appearance is now dominated by thick, grey hair and sunken, half-closed eyes.

Lupita grew up in Santiago, Chile, where she trained as a journalist. After the dictator Augusto Pinochet took power in a US-backed coup d'état in 1973, she worked for the resistance, publishing pamphlets about the regime's atrocities. Her colleagues were imprisoned, her father was tortured, and in 1978 she was evacuated by the United Nations and taken to the UK.

Her English wasn't good enough to continue work as a journalist, so after studying, she found a job as a social worker for Lambeth council. She enjoyed reading, painting, but more than anything she loved to travel. She reels off the countries she has visited – Scandinavia, India, China, Egypt, Ireland, Latin America. 'I loved to be there, with the people,' she says. 'Eating in the markets, feeling, seeing, being with their culture.' Then, over a period of six months, when she was 58, she went completely blind.

When Lupita was a child, she suffered from the infection toxoplasmosis, which made her shortsighted. The parasite had stayed dormant in her body, and now it destroyed her sight entirely. Lupita was divorced, and living alone. Completely self-reliant before she became blind, she now couldn't even make a sandwich – she ate by taking bread in one hand and cheese in the other.

'I was in shock. Sitting in this chair for a year,' she says. But gradually she restarted her life, getting to know her flat – every corner, every pipe – by touch. She stripped out unnecessary items – all the plants, her collection of traditional hats from around the world, even her favourite woven rug from Mexico, in case she tripped on it. She kept only a few treasured possessions including the framed poster on the wall behind her, precious even though she can no longer see it; a joyful pattern of stripes and splodges by the English painter Howard Hodgkin. It looks like the view through a window onto blue sky, I say. 'Exactly, my bedroom window!' she laughs.

Lupita has her independence back. She has learned to shop for herself, clean, bake bread, and even sew, if she has help threading the needle. But what still distresses her more than anything is her lack of social contact. Once her sight went, she realised that her hearing, because of the lifelong toxoplasmosis, was also very poor. Without being able to compensate using her eyes, she feels that her deafness separates her terribly from others. 'People relate beautifully to eyesight,' she says. 'But dealing with a person who cannot hear is very annoying.' She finds being isolated within a group even more painful than being alone, so she avoids social events, from her granddaughter's birthday party to the lectures and concerts she used to enjoy.

Her only outings are to the supermarket. 'I stay many days on my own and do nothing,' she says. She spends her time with taped audio books turned up loud; right now she's listening to Bruce Chatwin's *In Patagonia*. She appreciates our visit today and sees her son and his family each weekend, but says, 'Next week probably I will be on my own all day, every day. With food, with water, but completely on my own.'

I ask how that feels. Things hurt more, say if she traps her finger in the door, when there's no one around to share her pain. And small everyday problems – a drawer that won't open, or a visitor who's late – 'For me, it's like a drama.' She tries to deal with it by laughing at herself, singing songs like *What shall we do with the drunken sailor?* But loneliness changes your pattern of thoughts, she says. 'I worry about the most stupid things.'

Worst of all, she feels cut off – from people around her, and from events around the world. The pitch of her voice shoots up and she pulls a tissue from her sleeve. 'I feel completely on the side, looking in.' She hates that she struggles to hear the news, and when she does hear about problems elsewhere, 'You get so frustrated. The inability to do anything but pray.'

'To me, the universe is connection, it's communication,' she says. 'If you start to lose that, you start to die.'

***

There's a growing body of evidence that Lupita is right about that. The realisation that social connection literally keeps us alive began in the 1950s, when James House, an epidemiologist at the University of Michigan, dreamed up an ambitious project: to follow the health of an entire town.

House and his colleagues tracked the residents of Tecumseh, south-east Michigan, and in 1982 they reported an unsettling result. After adjusting for age and other risk factors, adults who reported fewer social relationships and activities were around twice as likely to die over the next decade.[6] Their lack of social bonds, it seemed, was killing them early.

Six years later, House and his colleagues wrote an analysis for the journal *Science*, reviewing the Tecumseh project but also subsequent studies of thousands of people living everywhere from Evans County, Georgia, to Gothenberg in Sweden, as well as lab tests and animal trials.[7] They concluded that social isolation is as dangerous for health as obesity, inactivity and smoking. The evidence was as strong as in the landmark US government report that in 1964 officially linked smoking with lung cancer.

House's paper had a dramatic impact. At a time when scientists were only just beginning to realise that the mind could affect health, the idea that our social life could be as important as physical factors such as diet or smoking was revolutionary. Since then, epidemiologists have compiled more evidence supporting House's finding. In 2010, US researchers analysed 148 studies following more than 308,000 people and concluded that lacking strong social bonds doubles the risk of death from all causes.[8] That confirms House's finding that in western societies, at least, social isolation is as harmful as drinking and smoking, and suggests that it is actually more dangerous than lack of exercise or obesity.

Of course, when we have social support, we live more healthily. We have someone to cook us meals, take us to the doctor, and nag us not to drink or smoke. This is a powerful effect, but the difference in death rates persists even after accounting for it. People who have warm relationships, rich social lives, and who feel like they are embedded in a group, 'don't get as sick, and they live longer,' says Charles Raison, a psychiatry professor and mind–body medicine researcher at the University of Wisconsin–Madison. 'It's probably the single most powerful behavioural finding in the world.'[9]

Back in 1988, when House and his colleagues published their landmark analysis, they warned that western society was changing in a way that could have dire consequences for health. Compared to the 1950s, they pointed out, US adults in the 1970s were less likely to belong to voluntary organisations, less likely to visit informally with others and more likely to live alone.

Rates of marriage and childbearing were falling too, meaning that the twenty-first century would see a steady increase in the number of older people without spouses or children. 'Just as we discover the importance of social relationships for health,' the researchers warned, 'their prevalence and availability may be declining.'

House's predictions were correct. Western society has continued to fragment. During the past two decades, the average household size in the US has shrunk. According to the 2011 census, 32 million people in the country now live alone; that's 27% of households, up from 17% in 1970.[10] When researchers asked a representative sample of Americans in 1985 how many confidantes they had, the most popular answer was three. When the study was repeated in 2004, the most popular answer – given by 25% of respondents – was none.[11]

***

When we're away from someone we love, we say that it hurts. You might think of this description as metaphorical; brain-scanning experiments suggest, however, that it's uncannily accurate.

It turns out that experiences of social exclusion or rejection – such as being shunned in a game, receiving negative social feedback, or viewing images of deceased loved ones – activate exactly the same regions of the brain as when we are in physical pain.[12] When we're socially rejected or isolated, we don't just feel sad. We feel injured and under threat.

Likewise, stress researchers have found that our bodies respond to social conflict – being criticised or rejected by others – in the same way we respond to imminent physical harm. It's no coincidence that one of people's most common fears is public speaking, or that one of the most effective tools psychologists have for triggering the fight-or-flight response, the Trier Social Stress Test, requires volunteers to perform in front of a panel of stone-faced judges. Carrying out similar tasks when nobody's watching is nowhere near as stressful.

The lack of social ties, although less acute, can be just as toxic over time: even if they score low on conventional measures of stress, lonely people have high baseline levels of stress hormones and inflammation, with all of the health problems that entails.[13] Social support also seems to shield us against difficult circumstances – those without it are much more susceptible to other stresses when they come.

But why do social rejection and isolation affect us so dramatically? Having no friends might not be pleasant, but it's hardly a matter of life and death. That's where I'm wrong, says John Cacioppo, a psychologist at the University of Chicago, Illinois, and probably the world expert on loneliness.[14]

He points out in his 2008 book, *Loneliness*, that for most of human history, becoming separated from others put us at imminent risk of starvation, predation or attack. Social isolation was indeed a death sentence, as much a threat to our survival as hunger, thirst or pain. As a result, we've evolved to be so desperate for human contact that if deprived of it we can even form attachments to inanimate objects, like Tom Hanks's character in the movie *Castaway*, who has a meaningful relationship with a volleyball he calls Wilson.

But you don't need to be marooned on a desert island to feel lonely. If we don't feel cared for, we can feel lonely even when surrounded by others: at college; on a crowded bus; in a strained marriage. After all, being among a hostile tribe is just as dangerous as being alone.

The impact of loneliness, then, depends not on how many physical contacts we have but how isolated we *feel*. You might have only one or two close friends, but if you feel satisfied and supported there's no need to worry about effects on your health, Cacioppo tells me. 'But if you're sitting there feeling threatened by others, feeling as if you are alone in the world, that's probably a reason to take steps.'[15]

Such 'loneliness in a crowd' is an increasing problem in modern society as we move around, often living far from family and friends. Studies in western countries suggest that 20–40% of adults are lonely at any one time, with one of the loneliest populations studied being college freshmen.[16] Most of us soon reach out to others or our circumstances change. But 5–7% of people report feeling intensely or persistently lonely.

One reason for their plight is that, like stress, chronic loneliness reshapes the brain, in this case making people more sensitive to social threat. Lonely people rate social interactions more negatively, are less trusting of others, and judge them more harshly. There's an evolutionary logic to this too: in a hostile social situation it is vital to be alert to betrayal and potential harm. But it can make lonely people reluctant to reach out to others. Feeling threatened also disrupts their social skills, says Cacioppo, leaving them focused on their own needs at the expense of anyone else's. 'When you talk to a lonely person you feel like they are feasting on you,' he says. 'Not in a good way.'

***

In 2007, Cacioppo published a result that opened a new window into how our physical make-up is influenced by the contents of our minds. He showed that stress – especially social stress – doesn't just affect the brain. It filters right down to our DNA.

From a group of 230 elderly Chicagoans, Cacioppo selected eight of the loneliest, who had felt isolated for several years, and six of the most connected, who reported that they had great friends and social support. He sent samples of their blood to molecular biologist Steve Cole at the University of California, Los Angeles, who analysed which genes were active in each group. The pattern of gene expression varies in different cell types, so Cole focused on the white blood cells of the immune system, because what these cells do – whether they cause inflammation or produce antibodies, for example – is crucial for health.

The Chicagoan's social world view had a dramatic effect on what was happening inside their cells.[17] Of about 22,000 genes in the genome, Cole found significant differences in more than 200 – which were either activated to produce more of a particular protein, or turned down to produce less. Individual genes might look different by chance; what was striking, says Cole, was the broader pattern.[18]

A large proportion of the lonely people's up-regulated genes were involved in inflammation, whereas many of their down-regulated genes had roles in antiviral responses and antibody production. In sociable people, the reverse was true – biological activity in their immune cells was skewed towards fighting viruses and tumour cells and away from producing inflammation. Crucially, the difference related most strongly not to the actual size of the volunteers' social networks but to how isolated they *felt* themselves to be. It was a very small study, but one of the first ever to link a state of mind with a broad, underlying change in gene expression.

The result suggests that our immune system is fine-tuned to respond to our social surroundings. It makes perfect sense that we evolved this way, says Cacioppo. In the past, people in a close-knit group would be at risk from viruses, which spread easily between individuals in close contact, or – because they would likely survive longer – from longer-term conditions such as cancer. An isolated person, by contrast, would have more to fear from physical attack, so their survival would depend on triggering branches of the immune system involved in wound healing and defence against bacterial infection. In today's world, however, this gene expression profile is a double whammy, increasing the risk of chronic inflammation-related conditions while leaving us more susceptible to viruses and cancer.

The researchers have since replicated that preliminary result in a larger sample,[19] and Cole has seen the same effect in other types of social adversity in humans and other primates, from macaques placed in unstable social groups, to people caring for dying spouses.[20]

Cole is now starting to test whether it's possible to reverse this adverse genetic profile. For example, a 2012 trial of 79 women recently diagnosed with breast cancer found that group stress-management therapy reduced expression of inflammation-related genes and pushed women back towards an anti-viral profile.[21] 'Our conclusion was that mood matters,' says Michael Antoni, of the University of Miami, Florida, who led the study.[22]

Not everyone agrees, in particular James Coyne, a health psychologist and emeritus professor at the University of Pennsylvania in Philadelphia and prominent sceptic of positive psychology. Particularly when it comes to cancer: researchers who claim that psychological factors can influence disease progression put pressure on patients, he argues, and risk blaming those who don't recover for not thinking in the right way or attending the right classes. 'They claim that if you make the right choices, you'll be healthy. And if you don't, you'll die.'[23]

Whether social support helps cancer patients to live longer has been controversial ever since Stanford psychologist David Spiegel found that group therapy doubled survival time in a 1989 trial of 86 women with metastatic breast cancer.[24] There have been plenty of attempts since to replicate that result, of which eight concluded that therapy does improve survival and seven have found no difference.[25] Results from epidemiological studies are mixed too, but in 2013, Harvard researchers who followed 734,000 patients found that for all the cancer types they looked at, people who were married were 20% less likely to die from their cancer, even after controlling for practical advantages such as help getting to appointments and taking medication on time.[26]

Overall, Spiegel claims the balance of evidence is in favour of there being a significant effect on survival,[27] whereas Coyne concludes that 'the whole idea that psychological factors can affect the lives of cancer patients is rubbish'. He describes Antoni's trials as too small to show anything useful and like researching the money you get from the tooth fairy: investigating a mechanism when it hasn't been established that there's an effect to be explained.[28]

'Everything we're doing is preliminary,' responds Antoni. 'We do need to be cautious. But each year, studies are showing results in a similar direction. They are showing that if we change the psychology, physiological changes do parallel that.' Antoni is now following 200 women for up to 15 years after receiving therapy, to see if this has any effect on cancer recurrence or survival time.

In general, the idea that social relationships influence gene expression in a way that's relevant for health is supported by insights from an emerging field called behavioural epigenetics. Epigenetics refers to a process in which DNA in a cell becomes physically modified, or tagged, in a way that controls long-term how genes in that cell are activated. This is what allows cells in our bodies to develop into different tissues – skin, nerves, white blood cells – even though they all contain the same DNA. Scientists used to think that once epigenetic tags were set in the embryo, they were fixed for life. But research now suggests that some of them at least can be altered later – and by social cues.

Some of the crucial experiments involve rats. When mothers nurture their young by licking and grooming them, the female pups grow up to be doting mothers themselves, with a healthy physiological response to stress. Meanwhile neglected pups grow up sexually promiscuous and hypersensitive to stress, and they ignore their own offspring. Researchers recently discovered why: when the pups are licked and groomed, it affects the epigenetic tagging of genes that encode receptors for the sex hormone oestrogen, and for the rat equivalent of cortisol.[29]

The same difference in the cortisol receptor gene shows up in human suicide victims who were abused during childhood[30] – a hint that similar processes may occur in humans too. Other studies show that patterns of epigenetic modification differ depending on socioeconomic circumstances, between institutionalised children and those raised by their biological parents, and even sometimes between identical twins.[31]

We've already heard how when children are exposed to adversity, their maturing brains become sensitised to stress. Epigenetics provides a second way in which early trauma – in particular a harsh social environment – can become programmed into our physiology, helping to explain why people brought up in tough environments later suffer from so much chronic disease. The research so far is preliminary – people are not the same as rats. But it's possible that adversity we experience in infancy (or in the womb) tags our genes in a way that subsequently raises inflammation levels and makes our immune system hypersensitive to threat.

New-age and holistic healers have seized on the concept of epigenetics as proof of what they claimed all along – that we can control our DNA and therefore heal ourselves using our minds.[32] Such claims are vastly exaggerated and misleading – researchers are only just starting to work out the balance between epigenetic changes fixed in childhood and those that remain fluid later in life. They're also not sure how early these changes occur (although extrapolating from animal studies suggests that we may be most susceptible before about age two). Pinning down the precise nature, mechanisms and timing of these changes – let alone their implications for health – is going to be a mind-boggling task.

But already it seems clear that we don't inherit from our parents a single 'biological self'.[33] Instead our genomes encode a wide variety of potential selves, and our social environment – including our perception of that environment – helps to determine which of those selves we become.

***

In her brick bungalow in Milledgeville, Georgia, 69-year-old Susan reaches for the bookshelf and takes down a large, glass jar filled with coloured cards. She picks some out to show me: they're a mix of simple chores and rewards, from 'wipe off kitchen cupboard doors' and 'dust furniture in one room' to 'go out for dinner' and 'extra TV time'. The jar is a memento of a pioneering trial she took part in, more than a decade ago.

The trial was run by the University of Georgia's Gene Brody. When he began studying impoverished families in rural black belt communities, he knew that their kids were at risk for behavioural problems such as alcohol abuse. But not all of them succumbed. So his first question was, why not?

He spent ten years studying thousands of families in places like Milledgeville, comparing kids who go off the rails with those who seem resilient to the stress of their surroundings. What allows some individuals to stay psychologically strong in such a tough environment? The one thing that protected the kids better than anything else, it turned out, was a particular style of parenting. Just like in the rats, the right care from a parent at this crucial stage in development protected them later in life.[34]

The most resilient kids were brought up by firm, vigilant parents – perhaps more strict than you'd find in populations living in less threatening environments. But crucially, these parents were also affectionate, communicative and highly engaged in their children's lives. Brody called it 'nurturant-involved' parenting. These kids knew where the boundaries were, and that there would be sanctions for bad behaviour. But they also knew this was because their parents loved and cared about them.

Brody designed a seven-week course to teach these principles to parents (and grandparents) attending with their 11-year-old sons and daughters. The course emphasised discipline as well as communication skills, with sessions on topics like 'supporting our youth' and 'making and enforcing rules'. He called it the Strong African American Families (SAAF) project. Then he ran a randomised controlled trial with nearly 700 families, to see what difference the course would make.[35]

Susan and her granddaughter Jessica were part of that original study. Susan says she already raised her kids and grandkids in a strict, loving way, but that she learned some useful tricks on Brody's course, like the reward jar. Whereas Jessica's older brother Kevin is in and out of jail, Jessica, now 24, did well in school and is at art college in Atlanta studying design and marketing. Susan proudly shows me one of Jessica's paintings on the wall – it's beautiful, showing two tall African women and a child, silhouetted against red earth, black hills and a yellow sky.

When he looked across all 700 families, Brody found that whereas parent–child relationships in the control group declined in the months following the course, in SAAF families, these relationships got stronger. This in turn improved behaviour: after five years, the SAAF kids were drinking only half as much as the controls.

But was there any lasting effect on physiology? To answer that question, Brody recently collaborated with Northwestern University's Greg Miller. The pair collected blood samples from nearly 300 of the families after eight years, when the kids were 19, and measured six different inflammation markers. Those in the SAAF group had significantly lower levels of every single one.[36] The effect was strongest for the most disadvantaged families, and it was mediated by changes in parenting: the more parents had shifted towards the nurturant-involved style, the lower their kids' inflammation.

It was a stunning result. Years later, long after these children had left home, this short intervention at age 11 was still dramatically affecting their biology. Miller and Brody are continuing to follow the trial participants, to see whether these differences in inflammation levels do indeed translate into health benefits as they age.

Across town from Susan's house, Monica and her teenage daughter Takisha have just finished the SAAF course when I visit. Monica says the classes have helped her to think about how to communicate with her daughter more positively, such as when Takisha said she wanted to be a singer. 'She really don't have a voice for singing,' says Monica. 'But I didn't realise I was tearing her down by saying that. This gave me another way to talk to her about the singing and not make her feel bad, to help her see that she has other options.'

Nurturant parenting might have reduced inflammation levels in the SAAF study by influencing health behaviours; that's something Miller wants to look at more. But the young people in the two groups didn't differ in body weight or rates of smoking. Instead, he thinks the communication skills training highlighted by Monica is at least part of the explanation. 'I suspect that it helped forge relationships and communication strategies between parents and their children that still are to this day a source of support for the kids.'[37]

Monica feels that it's too late to change her own situation, but hopes she can support Takisha to live a full life. 'I want her to have opportunities, to go see the world. I don't think that's too much to ask.'

The primary aim of the SAAF course is to help make that happen – showing children like Takisha how to build a strong self-image and resist peer pressure, and helping parents like Monica to support their kids through tough circumstances. If Takisha can stay out of trouble and do well in school now, she has a better chance of one day going to college and building a future career. But the results of Brody and Miller's trial suggest that strengthening Monica and Takisha's bond might do much more than that. By making Takisha more resilient to the biological effects of adversity, it might protect her against chronic disease for the rest of her life.

***

Brody's work shows that intervention in childhood might be able to halt the trajectory of stress sensitivity before it leads to chronic disease. But what if we miss that window? Seven hundred miles north of Milledgeville, researchers are working to strengthen social connections at the opposite end of life, among elderly residents of inner city Baltimore.

We've already heard how as we get older, the brain's prefrontal cortex, crucial for self-regulation, rational thinking and social relationships, starts to decline more rapidly than other parts of the brain – a process that is accelerated in people who are lonely or chronically stressed and that ultimately leads to dementia.[38] Michelle Carlson, a neuroscientist at Johns Hopkins Bloomberg School of Public Health, Maryland, was looking for ways to slow that decline. Old people tend to be isolated and sidelined, becoming less and less engaged in the community as they age. Carlson wondered what would happen if they were immersed instead in a rich social environment.

She worked with colleagues to develop a project called Experience Corps, in which elderly adults spend 15 hours a week volunteering in deprived elementary schools, helping kids to read. Most health interventions, such as exercise programmes, tend to have high drop-out rates even if they only take a few minutes each week. Fifteen hours was a 'crazy' amount of time to ask people to commit, says Carlson. Yet the volunteers stuck at it through the entire academic year. 'We tell them we need them, their wisdom and experience,' she says. 'They do it not for themselves but because the kids are waiting for them.'[39]

The volunteers formed close relationships with the children they were helping, creating a 'magic', according to Carlson, that isn't always there with teachers or parents. Many of the students are from troubled backgrounds, she says, but the older volunteers have the patience and experience to see beyond difficult behaviour to what children might be experiencing at home, at the same time as expecting them to succeed. 'They really can sometimes connect with the child on a different level.'

The programme significantly improved the academic achievement of the kids, but also the health of the volunteers. 'It was like watering them,' says Carlson. A pilot trial published in 2009 suggested that over a school year, the volunteers' activity levels increased and their legs got stronger – measures that normally decline with age.[40] They also performed better on cognitive tests, and had increased activity in the prefrontal cortex.

Carlson is now completing a two-year randomised controlled trial of the programme. She's still writing up the results, but so far has published a brain imaging study of 123 volunteers, focusing on the hippocampus (which works with the prefrontal cortex and is important for learning and memory).[41] The hippocampus usually shrinks with age and becomes impaired in the early stages of Alzheimer's. Yet in the volunteers, it got bigger. Age-related damage in their brains was being reversed.

Results like this suggest that we should see ageing differently, says Carlson. 'We over-estimate all the negatives about ageing, and we don't sufficiently emphasise what gets better with age. What gets better is that we have accumulated a lifetime of wisdom and knowledge. And we don't have a vehicle for giving it back.'

When we're old, just as when we're young, she argues, we still desperately want to have a purpose in society. Her comments make me think of Lupita, who has been active in politics and community her whole life. She's witty, brave, bursting with stories and experience, but is now forced to sit on the sidelines, unable to do anything but pray.

What if we reshaped care for the elderly not around managing their decline, but *harnessing their abilities*? We could 'use that ageing brain to give back to a society that's in great need,' says Carlson. The population is ageing, she points out; within 20 years we'll have more adults over 65 than children under 18. 'We don't know what the message does to a person when they are told ageing is a time of deterioration. If we reframe it, and say ageing is a time to give back to others, it might actually help them age better.'

***

Fhena is a large woman strikingly dressed in a voluminous mauve cloak. Her Afro hair has a splash of silver at the front, held back at the sides with black combs. She seems warm and happy, even radiant, and I tell her so.

You wouldn't have thought that a few months ago, she replies. Fhena has two sons: Ahav, who is five, and Analiel, three. Ahav talked early, but when he was about 18 months old he stopped. Other skills, such as catching balls and potty training, went too. And he became violent. 'It was beyond devastating,' she says. 'To see such promise early on, then see it disappear and not be able to go in and get it back.'

In 2012, shortly after his younger brother was born, Ahav was diagnosed with autism. Occupational and speech therapy helped dramatically, and Fhena was just beginning to accept the situation when Analiel regressed too. 'It was like having the same child twice.'

They would feed off each other, with up to ten intense melt-downs a day. 'I've had a cracked nose, busted lip, I have teeth marks on my arm,' Fhena says. 'I was getting two-to-three hours' sleep a night.' As with Lisa, the mother we met in Chapter 8, her marriage did not survive the pressure, so she was caring for the children alone and sometimes feared for her own safety. 'I've had it so that one of them is sitting on me holding me down and the other one is choking me.'

Fhena is a singer and performance artist from Atlanta, Georgia; she's naturally confident and gregarious. 'I've performed in Israel, Ghana, Antigua, all across the US,' she says. Before having kids, she performed live four to five times a week. She produced shows too, and released a CD, called *Beauty from Ashez*. But after her sons' diagnoses, all that stopped.

Without access to her beloved stage or studio, she felt trapped and hopeless. She also suffered from chest pains, headaches and insomnia. 'My body was in constant pain, I was walking like an old person. Some of it was because I was being punched and hit, but most of it was the stress lodged in my body.' Before the autism, she says she never took medication, not even during childbirth; now the first thing she reached for every morning was ibuprofen.

Then she took part in an experimental course being run at Atlanta's Marcus Autism Center, and it changed everything.

***

Brody's parenting course and Experience Corps are striking exam-ples of how strengthening social bonds within a community can improve people's lives and health. But can we take a more direct approach? What happens if we train ourselves to see the world in a more socially connected way?

The technique Fhena learned was developed at nearby Emory University, but has its origins in India. Its creator, Lobsang Negi, was born in a remote Himalayan village near the border with western Tibet. He trained as a Buddhist monk in southern India before being sent to the US in 1990 to set up a meditation centre in northern Georgia. He then relocated to Emory as a PhD student, and eventually took a faculty position in the university's religion department.

After a spate of suicides at Emory in 2003–04, a student came to Negi. She was concerned about mental health on campus and was impressed by some of the Buddhist principles Negi taught in his classes. Could he come up with an intervention that might help?

Negi came to the conclusion that what distressed, depressed people need most is a way to forge healthier relationships with those around them. Like Jon Kabat-Zinn, he took Buddhist principles and developed a secular course, but instead of focusing on mindfulness, Negi's course centres on compassion.

When I meet Negi in a restaurant close to Emory's campus, he's immaculately dressed in a pressed blue shirt and well-cut jacket and looks just like a western businessman, except for a string of amber-coloured prayer beads that peek out from the cuff of his jacket. He speaks with a soft voice and slight accent as he tucks into mushroom ravioli.

Cultivating compassion for others is more important than ever, he argues. Throughout most of human history, we've lived in relatively small groups. But now, 'We live in such a complex and ever-shrinking world. Each day we intersect with others who come with very different cultural, religious and socioeconomic backgrounds.' To cope with that shift, he believes that we must take the compassion we naturally feel for our loved ones, and learn to extend it even to those with whom we seem to have nothing in common.[42]

His course, called Cognitively-Based Compassion Training (CBCT),[43] involves meditating on feelings of love and kindness but also thinking carefully about how we might see the world in a new way. However different people may look, deep down we are all living beings who want to be happy. Reflecting on what we all share creates a sense of connection, says Negi, which makes it easier for us to respond to others' needs and difficulties.

The same is true for interdependence, 'the idea that we can't just survive by ourselves, with no help from others'. Even the simplest item we need to survive, like a sandwich, connects us with many other people, he points out – from farmers to supermarket workers. Extending that analysis to all of the things we need to get through a day – such as heating, electricity, roads, cars, fuel – demonstrates that we're dependent on a vast number of people.

If we spend some time thinking about all this, 'It's only natural for us to feel more grateful and more tenderly towards others,' Negi reckons. And that, he believes, is the foundation for healthy, meaningful social bonds. But does it work?

To find out, Negi teamed up with Charles Raison, an Emory psychiatrist (now at the University of Wisconsin–Madison), who studies the effects of inflammation on health. 'I was very interested in [whether] you could train people to see the world in a way that looked like your social connectivity was enhanced,' says Raison. 'I wanted to see if that would turn down inflammatory responses to stress.'

CBCT is generally taught as weekly sessions that include discussions, exercises and meditation, which attendees are also encouraged to practise at home. In the first trial, of 61 freshmen students, the course didn't significantly affect responses in the gruelling Trier Social Stress Test, compared to a group of controls. But among those who took the course, the more time they had spent meditating at home, the less distress they felt during the test and the smaller their inflammatory response.[44]

Raison and Negi found the same thing when CBCT was taught to abused teenagers in the Atlanta foster care system. Simply being exposed to the class didn't have a significant effect. But the more the kids practised, the bigger their reduction in stress hormones and inflammation.[45] There's some preliminary evidence that CBCT helps to improve empathy and social relationships too. In a small brain imaging study, students who took the course were more accurate at reading emotions from photographs of facial expressions, with more activity in the relevant region of the brain.[46]

The team has also taught CBCT to five-to-eight year olds in a local school – breaking the discussion principles down into games and stories. 'They got it faster than any adult group I've ever taught,' says instructor Brendan Ozawa-de Silva.[47] The results are not yet published, but Ozawa-de Silva says that after compassion training, children had more than twice as many friends as those in a class that were taught mindfulness. The course also helped to break down the divide between 'in groups' and 'out groups' – the CBCT children had more mutual friends and more cross-gender friendships. And they scored better on a story-completion task that assesses the ability to appreciate other's perspectives.

Larger trials are needed to confirm all of these results, and Negi and his colleagues are now studying the effects of CBCT in a range of communities at risk of stress, including Emory medical students, veterans with PTSD – and caregivers. For Fhena, the course, led by Marcus Autism Center psychologist Samuel Fernandez-Carriba, was a revelation. 'The fog started clearing,' she says.

During the course, Fhena says she realised that autism had come to define her children in her eyes. 'All you see is a burden. It was robbing me of so much I could give to them.' Instead of being overwhelmed by her own stress and misery, she started to view the world from her kids' perspective, and to see them as individuals in their own right. 'In the class, I released a feeling of entitlement,' she says. 'The feeling that I was supposed to have a life without these challenges.' She had always tried to be a good person. 'I thought, this isn't what I put into the pot, why am I getting this out?'

'Then I realised. These special beings were given to me *because* of what I put into the pot.'

And with that single thought, much of the stress in Fhena's life disintegrated. Instead of feeling bitter and resentful, she says, 'I'm enjoying being with them.' And her children have responded beautifully. 'Every day there is a new blossoming,' she says. 'Ahav is drawing cruiseships in 3D detail. Analiel is writing 25 songs a day.' And the best moment of all, when Ahav said, 'Mommy, I'm so proud of you. Because I know that you love me even more now.'

We've been speaking in Fernandez-Carriba's office at the Marcus centre, and Fhena takes the pair of us downstairs to meet her boys, who have just finished a behavioural therapy session. They're achingly cute, bouncing around with red anoraks and long, dark eyelashes. Analiel sings a song about a turtle and puts a green rubber band on my wrist. Ahav proudly shows me a red-and-blue transformer, swiftly morphing it into a truck. Then he turns to Fernandez-Carriba. 'Do you know how we hug in Hebrew?' he says, and gives the doctor a lopsided, one-armed cuddle.

11

GOING ELECTRIC
Nerves That Cure

This is no ordinary medical practice. I'm in a rambling farmhouse set among the frosty fields of Chard, Somerset. The consultation room is yellow and spacious, with sloping ceilings, a comfy sofa and a tall vase of fresh flowers. As I look out of the huge, triangular window, a horse trots past.

Patricia Saintey – petite, strawberry blonde, with a peach frilly cardie – clips a monitor onto my ear. It will monitor my pulse by detecting blood flow, she explains. 'Now I'll pop you onto the biofeedback.'

The computer screen promptly shows a black line: my heart rate. Although our hearts speed up when we're stressed or when we exercise, I've always thought of my resting pulse as stable, making regular beats like a metronome. Now I discover that it constantly jumps around. Rather than a straight line, the graph shows a chaotic series of spikes, some large, some small. The amount my heart rate fluctuates, explains Saintey, is called 'heart rate variability' or HRV.

'You want to see if you can transform that jagged variation and make it into a coherent wave,' she says. A broad blue bar pops up on the left hand side of the screen. It's slowly pumping up and down, like a cylinder of water that fills and then empties.

Saintey asks me to breathe in time with the blue bar – five seconds in as it fills, five seconds out as it drains.

Then something striking happens. Within a few seconds, the difference between my lowest and highest heart rate is much larger than before – varying from about 60 to 90 beats per minute. And the line on the graph transforms from ugly random spikes into a smooth, snake-like curve.

Saintey works in Somerset as a part-time GP, but she also runs this private alternative health practice out of her home. She calls it Heartfelt Consulting, and it is based around a technique called HRV biofeedback. The idea is that you use the heart rate monitor and computer display to practise getting your heart rate into this smooth curve, a state described as 'resonance' or 'coherence'. Once you've got that nailed, try to increase the height of the wave: the difference between lowest and highest heart rate. By practising every day, says Saintey, we can learn to increase our heart rate variability and achieve this coherent state more often.

Proponents claim that this training has a range of benefits, strengthening our hearts, reducing stress, and even making us happier and more alert. Although Saintey offers the technique in the clinic, there's a growing range of portable devices that people can use to practise HRV biofeedback at home, from the FDA-regulated 'StressEraser' to the 'Inner Balance' sensor, sold by the Institute of HeartMath, which works with a smart phone and claims to 'reduce the negative effects of stress, improve relaxation and build resilience with just a few minutes of daily use'.

As a scientist, I like the idea of having an instant readout of what's happening in my body. And the change I see on the computer screen is intriguing – by choosing to breathe more slowly, I've caused my heart to beat in a dramatically different pattern. But these wide-ranging claims for benefits ring alarm bells. It seems unlikely to me that this simple exercise would have such potent effects. Indeed, HRV biofeedback has been criticised by Steven Novella, a clinical neurologist at Yale University School of Medicine and a prominent sceptic of alternative medicine, as nothing but 'bad tracings, technical artifacts and noise'.[1] This smooth curve might look pretty, but I'm not convinced it can really improve health.

It turns out I'm in for a surprise. Investigating heart rate variability leads me much further than I expected, to another crucial link between the mind and the body; research that might challenge our reliance on chemical drugs; and a baby girl called Janice.

***

3 May 1985 started just like any other Friday. Cecilia was preparing spaghetti in the kitchen of her third floor apartment in Brooklyn, New York, while her 11-month-old granddaughter, Janice, played happily on the floor. It was half past five, and Janice's parents would soon be home from work.

Then came a split second that changed everything. When the spaghetti was cooked, Cecilia grabbed the heavy pan and turned towards the sink to drain it. But the baby had stopped just behind her feet. She tripped and dropped the pan, pouring its boiling contents all over her precious grandchild.

One of the doctors called to treat Janice when she arrived at New York Hospital was 27-year-old Kevin Tracey.[2] It was his second year as a doctor, and he was training to be a surgeon. Although Tracey was used to seeing horrific injuries – gunshot wounds, head injuries – he was shocked by the sight of this tiny blonde girl with blistered, oozing skin. Her face was spared but deep burns covered more than 75% of her body, including her back, arms and legs.

Trying to numb himself to her pain, he stripped her clothes and covered her in antibiotic cream – without intact skin, dehydration and infection are huge risks – and estimated that she had a 25% chance of survival. Then he transferred her upstairs, to a steel-barred crib on the burns unit.

There, Janice endured a gruelling catalogue of interventions and treatments. Unable to eat, she was fed through a tube. She suffered daily sessions of agonising wound care, just like the burns patients we met in Chapter 6. Then there were several rounds of major surgery to cut away the burned areas and cover them with skin grafts – at first shaved from her unburned buttocks, and when that ran out, from cadavers.

She had a couple of crises. On Tuesday 7 May, her blood pressure suddenly plummeted and she fell into a coma: a phenomenon known as septic shock. Without sufficient blood pressure, the heart can no longer effectively pump blood around the body. Deprived of oxygen and nutrients, cells and organs die. In up to half of cases, septic shock is fatal.[3]

At the time, doctors thought septic shock was caused by toxins from bacterial infection. But often, as in Janice's case, no infecting bug is ever found. Tracey and his colleagues pumped gallons of intravenous fluid into Janice to try to raise her blood pressure, and infused adrenaline to boost her heartbeat and constrict her arteries. By Wednesday, however, Janice's hands and feet were turning grey, and her lungs and kidneys were starting to fail.

On Thursday morning, the crisis was suddenly over; Janice recovered as quickly and mysteriously as she had succumbed. But on Sunday 12 May, she developed another complication.

Tracey describes Janice's new problem, severe sepsis, as the 'pestilence of the 21st century'.[4] It is one of the most common causes of death worldwide, killing nearly a quarter of a million people a year in the US alone. It often affects patients who are already ill – with burns injuries like Janice, for example, or heart disease, cancer, infections or trauma.

In the 1980s, doctors assumed that severe sepsis too was caused by toxins produced by invading bacteria. It develops more slowly than septic shock. Patients show signs of infection and inflammation throughout the body, and gradually their organs stop working. This time, tests did show microbes in Janice's bloodstream. She developed a 104-degree fever. Then her kidneys, gut, lungs and liver all began to fail.

Antibiotics cleared the bacteria from Janice's blood but her condition didn't improve. She was on life support for days, with her family (who were only allowed to see her during brief visiting hours) keeping a desperate vigil by the elevators.

Once more, amazingly, this tiny child bounced back. By 28 May, her first birthday, it seemed for the first time that she was going to make it. Janice looked healthier than at any time since the tragic accident. She had drunk her first milk, and her burns were starting to heal. They had a party; Tracey recalls chocolate cake, streamers, and Janice laughing, with rosy cheeks. Everyone – her family and the entire medical team – was celebrating not just Janice's birthday but her miraculous recovery, her precious life. Just one more round of relatively minor surgery, and she could go home.

The next day, a nurse was feeding Janice a bottle of milk when the child's eyes rolled back in her head and her heart stopped. Tracey and a colleague carried out CPR, injected adrenaline, and repeatedly shocked Janice with defibrillators. They kept it up for 85 minutes. They even inserted an electrical pacemaker. But her heart did not restart.

When he was five, Tracey's mother had died of a brain tumour, and after the funeral the young boy asked his grandfather, a paediatrician, why surgeons couldn't just cut the tumour out. The tumour sends projections into the surrounding tissue, the man replied. It wasn't possible to remove it without destroying the healthy brain too.

The five-year-old said that when he grew up, he was going to do medical research – he would find better techniques so that next time doctors would not have to stand by and watch a person die. Yet now, 22 years later, he was forced into exactly the same position with Janice. There had been nothing he could do.

Unable to speak even to pronounce the time of death, Tracey

walked out of the room. He did not see Janice's body, or her family, again. But the case haunted him. He suffered recurring nightmares, reliving her story yet each time with the awful knowledge of how it was going to end.

Tracey tells Janice's story in his 2007 book, *Fatal Sequence*. In the book, he says that when Janice died, he was due to start two years of research and hadn't been sure what his project should be, but now he knew. 'Janice's story compelled me to study sepsis,' he writes.[5] He wanted to understand what went wrong in Janice, and how it could be fixed.

His research would ultimately lead him to the same structure in the body as that targeted by HRV biofeedback: a meandering bundle of fibres called the vagus nerve.

***

Paul Lehrer, a psychiatry professor at Rutgers University in New Jersey, has dedicated his career to studying biofeedback. He wasn't convinced by its benefits at first, but then he saw a group of Russian children playing an intriguing computer game.

There are lots of different types of biofeedback, and the general idea is that by monitoring different aspects of our physiology in real time, we can learn how to shift our bodies into particular desired states – for example a state of relaxation. Lehrer studied electromyograph (EMG) biofeedback, which monitors muscle tension, for example, and finger temperature biofeedback, which is based on the fact that when we're relaxed, our extremities, including our fingertips, get warmer. They worked, but didn't seem to be more effective than more direct methods to relax the body, such as progressive muscle relaxation (a technique that involves tensing and then relaxing different muscle groups in turn).

Then in 1992, Lehrer visited St Petersburg, Russia, where his son was studying. While there he asked around to see if anyone was studying biofeedback, and was directed to a private clinic treating children with asthma. Staff at the clinic were using computer games to help the children increase their HRV. 'The best one involved a paint brush painting a fence that was filled with apparently rather funny Russian graffiti,' recalls Lehrer. 'If the amplitude of heart rate fluctuations was high enough, the fence was completely painted. If not, part of the fence was missed.'[6]

It was intriguing, but Lehrer had no idea if or how boosting HRV might work, for asthma patients or anyone else. A couple of years later, Lehrer visited St Petersburg again, and was introduced to a physiologist and engineer called Evgeny Vaschillo, who had studied HRV biofeedback in Russian cosmonauts. Vaschillo showed the cosmonauts a sine-wave pattern on an oscilloscope, and asked them to match it with their heart rate. With practise, the cosmonauts achieved huge fluctuations of up to 60 bpm.

Lehrer helped Vaschillo to get his work published in the US,[7] but not before the paper was rejected by various physiology journals. One reviewer objected that such a huge variation in heart rate simply isn't possible. Either the data were inaccurate or faked, or Vaschillo was studying 'some kind of yogis'.[8] In fact, what was happening to the cosmonauts' hearts was a simple physical phenomenon: something that Vaschillo, with his engineering background, recognised, but that the physiologists had missed.

There are several processes in the body that cause our heart rate to fluctuate. One is the 'baroreflex'. Reflexes controlled by the nervous system monitor conditions in the body and act to keep us safe, without requiring any conscious thought. Some affect our behaviour; if you touch something hot, for example, a reflex causes you to pull back your hand. Others constantly adjust various aspects of our physiology to keep them within safe limits.

The baroreflex does this for blood pressure. It's controlled by stretch receptors in artery walls. If blood pressure goes up, this activates the stretch receptors, sending a signal to the brain stem, which then sends a signal back to slow the heart so that blood pressure falls. If blood pressure falls too low, the stretch receptors send the opposite signal, and our heart rate goes up again.

A second process that varies our heart rate is called 'respiratory sinus arrhythmia' (RSA). When we breathe out, our heart rate falls slightly, bouncing back up again when we breathe in. This maximises oxygen transfer around the body when we have a lungful of fresh air, while slowing the heart and allowing it to rest as we exhale.

Both forms of variability are essential for a healthy, resilient heart; people with low HRV are much more likely to die of heart disease.[9] This is partly because having a more sensitive baroreflex (defined as a greater change in heart rate for each shift in blood pressure) makes us better able to recover from changes in blood pressure, like those we experience during stress or exercise. And if the heart fails to slow when we breathe out, our overall heart rate is higher. This strains the heart, increasing the risk of hypertension, stroke and other cardiovascular problems.

Usually these two patterns of heart rate variation happen on different timescales. RSA causes the heart rate to go up and down as we breathe, while the baroreflex is slower, taking about five seconds each way. When the two are superimposed, we get an irregular, jumpy pattern.

But if we slow our breathing down to match the baroreflex – five seconds in, five seconds out – the two patterns occur on the same time scale, and their peaks and troughs become superimposed, making a single smooth wave. And if we get it just right (the exact speed depends on how big you are and how much blood you have), this leads to a phenomenon known to engineers as 'resonance'. Each time the baroreflex goes up or down, the extra variation from the RSA gives it a little kick at precisely the right moment – like pushing a playground swing – causing the fluctuation in heart rate to become bigger and bigger.

Lehrer believes that this provides a beneficial workout for the heart and the baroreflex, making them more resilient.[10] Supporting this idea, there's some evidence that biofeedback improves HRV over time, even after the treatment is over, and that it helps to lower blood pressure.[11] Trials have also found benefits for pain, anxiety and depression, however, suggesting that the effects of HRV biofeedback aren't limited to the heart.[12] So why would changing our pattern of heartbeats affect our emotional state?

***

In the 1960s, a Harvard cardiologist called Herbert Benson was studying blood pressure in monkeys when a group of practitioners of transcendental meditation (TM) turned up at the medical school. They believed that they could lower their blood pressure just by meditating and they wanted the professor to study them. Benson didn't initially want to get involved in such a 'far out' practice,[13] but they persisted, and Benson was intrigued by their apparent abilities. So he turned his attention from monkeys to meditation.

In fact their blood pressure didn't change – the meditators had low blood pressure all the time (although Benson found in future experiments that TM did lower blood pressure in patients with hypertension).[14] But he was surprised to find that by meditating, the TM devotees could induce an ultra chilled-out state in which their breathing and metabolism slowed and their heart rate dropped.[15] Benson called it the relaxation response.

This response, it turns out, is the opposite of fight-or-flight. Whereas the arousal of fight-or-flight is triggered by the sympathetic nervous system, the relaxation response is orchestrated by an opposing nerve network called the parasympathetic nervous system. It's the parasympathetic system that calms us down after an emergency, pushing the balance back towards the non-urgent activities – digestion, sex, growth and repair – that we engage in when we're safe and at rest.

The main component of the parasympathetic nervous system

is the vagus nerve. From the brainstem it wanders down through the neck and torso, with branches that lead to various major organs, including the lungs, gut, kidney and spleen. One of its jobs is to act as a brake on the heart. The stronger the activity of the vagus nerve (described as 'vagal tone'), the more our heart rate slows during the baroreflex and as we exhale – and after stress – and the greater our heart rate variability. In fact, HRV is often used as a measure of vagal tone, and an indicator of how active our parasympathetic nervous system is in general.

As well as turning down stress throughout the body when we perceive that a threat has passed, the vagus nerve relays messages from the body back to the brain (in fact around 80% of its fibres run in this direction). Brain imaging studies show that people with high HRV also have more flexible and adaptive emotional responses to stress, whereas those with low HRV are hypervigilant, seeing even small problems as significantly stressful.[16] People with high HRV tend to have better working memory and can focus their attention better, and they are better able to regulate their own emotions and facial expressions.

Some studies even suggest that people with high HRV form stronger social relationships, and get more pleasure from social interactions. By contrast, people with low resting HRV aren't just at risk of heart disease. They are also more likely to have a range of psychiatric disorders including anxiety, schizophrenia and depression.

'HRV is important not so much for what it tells us about the state of the heart,' writes Julian Thayer, a psychologist and expert on HRV at Ohio State University, Columbus, but 'for what it tells us about the state of the brain'.[17]

When we slow our breathing to push up HRV, this stimulates the vagus nerve, which in turn tells the brain to switch off fight-or-flight. Biofeedback and meditation (and possibly other activities such as yoga and tai chi, which encourage slow, controlled breathing) probably have a similar effect. When biofeedback researcher Lehrer studied a group of Zen monks, he found that they were indeed creating a strong state of resonance.[18]

He argues though that because the speed of breathing needed to achieve resonance is slightly different for each person, maximising the effect with meditation alone can take years of practise, whereas with biofeedback, we can learn it in a few minutes. 'Most people are able to pick it up right away,' he tells me. 'That's very different from living in a Zen monastery for ten years!'

Whether all this translates into significant health effects long-term, however, is still up for debate. Lehrer points to clinical trials showing that HRV biofeedback helps with stress-related conditions from high blood pressure to asthma.[19] But the studies are generally small and haven't been well-assessed in meta-analyses.

'Unfortunately we don't have big drug companies out there supporting research on 20,000 people for each condition, so I can't say it works the same way penicillin works for infection,' admits Lehrer. 'The problem is that no one can make money doing this. Biofeedback equipment is easy to copy and cheap to make.' Even so, he characterises the evidence as 'pretty good'. Plus, he says, 'It's a nondrug treatment with very powerful effects. It's easy to learn. Why isn't everybody doing it?'

Lehrer appears to have hit the impasse suffered by many mind–body therapies – with nothing to sell, there's limited funding for research. But thanks to Kevin Tracey's work, interest in the vagus nerve is now exploding.

***

In 1985, when Tracey started working on sepsis and septic shock, doctors believed that these conditions were caused by invading bacteria. But, mysteriously, there were often no detectable pathogens. It hadn't occurred to anyone that devastating symptoms like those suffered by Janice could be created instead by our own bodies.

Scientists used to assume that any damage done when we have an infection was caused by the infecting organism. Slowly they realised, however, that many of the symptoms we suffer when we're ill – fever, weight loss, tissue damage, even fatigue and depression – are triggered not by pathogens but by our own immune systems, mediated by messenger proteins called cytokines.

Sometimes these symptoms are a necessary byproduct of the body's attempt to tackle infection. The raised temperature we experience during a fever helps to kill off invaders. Fatigue and depression encourage us to rest while we're ill, and to stay away from others so we don't spread the infection. Inflammation is crucial for fighting bacteria and clearing damaged cells.

But our bodies can get the balance wrong. Children, especially, can suffer dangerous seizures if their fever spikes too high. Sometimes the fatigue triggered by an infection never lifts. And Tracey showed that the acute septic shock suffered by Janice is caused when the body produces excessive amounts of a cytokine called TNF.

In a crucial experiment, he injected TNF into a rat; despite there being no infecting bacteria, the animal went into profound shock, its blood pressure fell catastrophically, and it died.[20] Instead of triggering an appropriate and proportionate inflammatory response, Tracey found, too-high levels of TNF essentially activate every white blood cell in the body. These clog up blood vessels, blocking blood flow and starving cells downstream of oxygen and nutrients. In other experiments, he discovered that severe sepsis – Janice's second crisis – is caused when a different cytokine, called HMGB–1, rages out of control.[21]

Tracey realised that these cytokines can cause other problems too. If TNF storms through the whole body, we suffer from acute shock. But if confined to particular locations it causes other inflammatory conditions – too much TNF in the joints contributes to rheumatoid arthritis; in the gut it can cause Crohn's disease. This insight led to a new class of drugs designed to inhibit or neutralise cytokines, including anti-TNF, which has since been used successfully to treat millions of patients.

It still wasn't clear why the body would unleash damaging amounts of these cytokines. Then in the early 1990s, while working at North Shore University Hospital in Manhasset, Long Island, Tracey made another revolutionary discovery. His team was working on an experimental drug called CNI–1493, which blocked production of TNF and other cytokines by white blood cells.

Tracey wanted to see if the drug could help treat stroke in rats. Ischaemic stroke causes brain damage when blood flow is blocked to a region of the brain. That damage is made worse when the dying cells release TNF. One series of experiments involved trying to prevent this by injecting a tiny amount of CNI–1493 directly into the brain.

But one day, CNI–1493 was accidentally injected into the brains of rats with a different condition. These rats had endotox-aemia, in which bacterial toxins cause very high levels of TNF to be released into the bloodstream, triggering septic shock. To Tracey's surprise, the tiny dose of drug in the rats' brains shut off TNF production throughout their bodies.[22] It was 300,000 times more effective than injecting the drug into a vein.

A signal must have been sent to the immune system, telling it to stop producing TNF. Far from simply responding to conditions in the body as had been assumed, the inflammatory response was being tightly orchestrated by the rats' brains.

How did the message get through? Tracey couldn't find any hormone released into the bloodstream. Then he had a radical idea – maybe it wasn't a chemical signal but an electrical one. He had seen work by another researcher, Linda Watkins at the University of Colorado, Boulder, in which she triggered fever in rats by injecting a cytokine called IL–1. She found that she could block the phenomenon by cutting the vagus nerve.[23]

We heard in Chapter 3 how Robert Ader and David Felten

first discovered that the brain and the immune system communicate via nerves. Watkins' experiment was further proof of the link, although this time the signal was carried not by the sympathetic nervous system, which Felten and Ader had studied, but by the parasympathetic system, and in particular the vagus nerve.

In Watkins' experiment, the signal travelled from the immune system to the brain; Tracey wondered if the vagus nerve could also carry messages in the other direction. Perhaps this was how a tiny dose of drug in the brain blocked TNF production throughout the body. In May 1998, he came up with a way to test the idea. He went to the operating room of the hospital and borrowed a handheld, battery-operated nerve stimulator.

Again, his experimental subjects were rats with endotoxaemia. They normally die from septic shock, but when Tracey stimulated the animals' vagus nerves with a pulse of electricity, their TNF production was dramatically reduced.[24] His makeshift treatment stopped septic shock in its tracks.

It was proof that as well as slowing the heart, the vagus nerve can act as a powerful brake on inflammation. Tracey called this the 'inflammatory reflex'.[25] Just as the baroreflex keeps blood pressure within safe limits, the inflammatory reflex protects us from the lethal weapons of the immune system. Rather than acting autonomously as scientists had thought for so long, the immune system communicates with the brain, which acts as a 'master controller'. If the brain detects a signal via the vagus nerve that inflammation has been activated in the body, it swiftly fires a return signal to calm it down again.

At last, Tracey could make a good guess as to what went wrong with Janice. Her nervous system must have failed as a result of her injuries – with either insufficient activity in the vagus nerve itself, or a problem further upstream in the brain. During her first episode of acute shock, the vagus nerve did not relay the signals needed to prevent catastrophic release of TNF. In the second crisis, sepsis, the vagus failed to block a flood of HMGB–1.

Despite each apparent recovery, the accumulated damage to Janice's organs was presumably too great for her to survive.

It seemed a fair bet that insufficient vagus activity is also behind many other conditions in which inflammation rages out of control. At lunchtime, Tracey drew a sketch on the back of a napkin – showing a person with an implanted pacemaker connected to an electrode on their vagus nerve.[26] A pulse of electricity had just saved his rats. Could it do the same for people too?

***

Altering our breathing rate may not be the only way to boost vagal tone voluntarily. HRV biofeedback appears to have a 'bottom-up' effect on the parasympathetic nervous system – altering heart rate in a way that stimulates the vagus nerve and in turn influences the brain. But experiments carried out by psychologists at the University of North Carolina in Chapel Hill suggest we can increase vagal tone from the top down, too, by changing our pattern of thoughts.

In a 2010 study, Bethany Kok and Barbara Fredrickson asked 73 volunteers to write down each day how happy they were and how socially connected they felt.[27] Over nine weeks, the volunteers' emotional wellbeing significantly increased – and so did their vagal tone.

The pair tested this phenomenon further in a randomised controlled trial published in 2013. Participants were asked to rate their emotions in the same way, but also to do daily loving kind-ness meditation (a practice that is similar to, but not the same as, compassion meditation). The same thing happened – after two months the meditation group felt significantly happier and more socially connected than the control group.[28] That emotional shift in turn increased vagal tone.

In both studies, those who had the highest vagal tone to start with benefited the most. Kok (now at the Max Planck Institute for Human Cognitive and Brain Sciences in Leipzig, Germany) suggests that reflecting on positive emotions triggered an 'upward spiral' mediated by the vagus nerve, in which body and mind influence each other in both directions. Positive emotions improved vagal tone, which in turn improved the volunteers' wellbeing even further. In a third, as yet unpublished, study, Kok devised a stricter test, in which volunteers simply rated the closeness of their three most meaningful social interactions, each day for 12 weeks. Those in a control group were asked instead to rate the usefulness of the three longest activities they had engaged in that day.

Vagal tone significantly increased in the social closeness group compared to the controls. 'What I keep finding,' says Kok, 'is that it's not just positive emotions that are important for the vagus, it's *social* positive emotions. If these positive emotions are not social, if they are not linked to feeling love and closeness and gratitude and all these things, then you don't get these relationships.'[29] We saw in Chapter 10 how social bonds can defuse our response to stress – it may be that this works at least partly via effects on vagal tone.

There are claims that biofeedback, too, works better if you try to think loving thoughts. A non-profit organisation called the Institute of HeartMath, based in Boulder Creek, California, promotes HRV biofeedback techniques that it claims are based on scientific research and used by hospitals, government agencies and companies around the world, as well as hundreds of thousands of individuals (Saintey's methods are ultimately based on principles developed by the institute). HeartMath techniques differ from the HRV biofeedback studied by Lehrer in that as well as breathing at the appropriate speed to create resonance, you also have to generate a 'heartfelt positive emotional state'. According to the institute's website, 'this emotional shift is a key element of the techniques' effectiveness'.[30]

Other claims made by experts at the Institute of HeartMath are downright nonsensical – they include that your HRV is directly connected to the earth's magnetic field and the electrical activity of the sun, and that your heart is capable of telepathically detecting information about events that haven't happened yet[31] – and their methods are often criticised as scientifically suspect.[32] After talking to Kok about her research, however, I wonder if they are right about the importance of positive emotions.

Sitting in Saintey's consulting room in Somerset, I decide to test the idea. During my biofeedback session, I first think about my children. I imagine hugging them tight until I'm so full of love it feels as though my heart will burst. My heart rate dutifully forms a lovely, smooth curve on the computer screen. Then, I attempt to will myself into a state of panic.

While keeping my breathing slow in time with the blue bar on the screen, I imagine tarantulas creeping up my arms, maggots crawling on my skin, an axe murderer behind my chair about to bring down his glinting blade. I focus urgent hate on the axe murderer. I feel a sudden burst of energy, with sharpened awareness and the rush of adrenaline flowing through my veins. But my parasympathetic nervous system is apparently not affected in the slightest. The smooth curve is uninterrupted, and my HRV actually goes up.

Biofeedback researcher Lehrer acknowledges that generating loving emotions might in theory have an effect on HRV longterm. 'But my pretty strong feeling is that whatever contributions your emotional state has while you're doing the technique is very small and almost impossible to see compared to the huge effect of breathing.' A couple of studies have compared HRV biofeedback with and without the heartfelt emotion that HeartMath techniques involve, he says. 'And they found absolutely no differences.'

Kok doesn't recommend reflecting on social closeness as a way to improve physical health either. The effect in her studies is statistically significant and important from a scientific point of view, she says, because it shows that in principle it's possible to use our thoughts to influence HRV. The impact is probably too small to have any clinically meaningful effect on health, however. She hopes that in future it will be possible to design effective ways to optimise HRV that have a psychological component. But for now, if you want to boost vagal tone, she recommends physical methods such as aerobic exercise, shown in various studies to increase HRV (fish oil supplements seem to work too).[33] 'That is going to get you the biggest effect the fastest.'

***

I'm sitting at a giant wooden dining table. A boisterous black puppy is sparring with an unimpressed cat, and Saintey is making lunch on the Aga. So what causes a conventional GP to build a practice based on biofeedback?

Saintey was a doctor in the army for ten years, she tells me, serving in places like Northern Ireland. Then she fell while skiing and ruptured the ligaments in her knee. She was discharged with a pension, and became a full-time GP in Somerset.

Working long hours, seeing 35–40 patients a day for ten minutes each, she felt stressed and increasingly disillusioned. She couldn't care for her patients as she wanted, and was losing faith in herself as a doctor. She felt that in too many cases, she simply prescribed pills and sent patients home, while ignoring the underlying issues – stress, abuse – that were causing them to come back again and again.

She ignored the lump in her breast too at first. She'd had a lump before that turned out to be benign, so she assumed this would be the same. But it was malignant, and by the time she got it checked out it had spread to her lymph nodes. She had surgery, followed by chemotherapy and radiotherapy. She was 42.

Supported by the payout from her health insurance, Saintey resigned from her job and took three years off. Her overwhelming emotion was relief that she was no longer working. She remembers one morning after her shower writing a message in the bathroom mirror: *I'm glad to be alive*. She decided to use her unplanned career break to explore how she might help her patients to live more healthily, rather than just tackling symptoms after they arose. She took a course in alternative medicine, and discovered biofeedback.

When her insurance payments stopped she returned to work as a GP part-time. Now she sees just 12 patients a day, three days a week, and stays late so she can give them 15-minute slots. 'I take a more holistic approach than a lot of my GP colleagues,' she says. 'I talk about the lifestyle aspects of keeping yourself well.'

Spending more time with her patients is crucial, she says. 'You can't start asking someone to make lifestyle changes that they probably have struggled with all their life, if you don't know them.' And in 2012, she started Heartfelt Consulting.

One of her patients is a 65-year-old grandmother named Carol, who worked as a pathologist and a nurse before retiring at 55 to do a history degree. Carol tells me that she had always been active and healthy, but when she was 60, while studying for her exams, she had a few panic attacks during which her heart would race. At the time she was drinking 'roughly 10 espressos a day' and wondered if that might be the cause, so she phoned her local surgery to check their guidelines for healthy caffeine consumption.

'Suddenly from one phone call I'm on this medical treadmill,' she says. She went through a battery of tests on her heart, including an ECG (which involved wearing a heart monitor for three days), an echocardiogram (in which her heart was examined using ultra-sound) and an exercise stress test. Everything was normal, except that she failed the exercise test.

Carol feels that the doctors ignored factors such as her caffeine consumption, and how anxious the tests were making her. Instead, the episodes of racing pulse she had experienced were diagnosed as paroxysmal atrial fibrillation (intermittent episodes of irregular heartbeat) and she was put on a powerful drug called flecainide, which slows the transmission of electrical signals in the heart.

The diagnosis had a huge impact on Carol. 'Suddenly from being a very well person I thought, "I'm ill, I'm going to be on medication for life."' She was just starting to take care of her new grandson as her daughter returned to work. 'I thought, "Gosh, I'm going to be in charge of this tiny baby. And I've got a heart problem! We live out in the depths of the country."'

Carol never did suffer from any episodes of atrial fibrillation, so over the next couple of years she persuaded her doctors to reduce her dose of flecainide, until eventually she was allowed simply to carry it with her, in case she needed it. But her anxiety persisted. 'I'd lost all confidence that I was fit and well,' she says. If she went away for the weekend, she checked where the nearest hospital was, in case she had an attack. If she went out walking, she made sure that her phone was always on. She avoided going to the theatre or cinema, in case she fell ill and needed to be carted out.

Then she went to see Saintey. For six months, she did fortnightly sessions of biofeedback at Heartfelt Consulting, and practised daily at home. Whereas she feels that conventional medicine simply contributed to her anxiety, she now appreciated being able to talk to Saintey about her concerns. And the biofeedback was 'wonderfully reassuring', she says. 'I could see that my heart was working well. I got more confident, and thought, I'm alright.'

Since taking the course, Carol says she's free of panic attacks. Now, if she feels herself getting anxious – when driving, in a crowded place, or waiting to see the doctor or dentist – she uses the breathing technique to calm herself down. What's more, her blood pressure, resting pulse and cholesterol levels have all dropped – without the use of drugs.

'The pharmaceutical industry has made a lot of patients very dependent on the system,' comments Saintey. 'We should be making them independent of the system.' Give people skills, she says, and make them responsible for their own health.

***

Monique Robroek is curvy and smiley, in her late thirties perhaps, with feathery hair and a drapey green blouse. She sits on the edge of a hospital bed at the Academic Medical Center in Amsterdam and pulls down the neckline of her top to show off a pink scar; a horizontal line a couple of inches long. Beneath it, she explains, is an implant like a pacemaker, with a wire that leads to her vagus nerve.

She takes a small black magnet – the size and shape of a car key – and swipes it across her chest, as if she's scanning groceries at the supermarket. The magnet triggers her implant to zap the vagus nerve in her neck with mild electric shocks. As she talks, her voice starts warbling. 'I get a vibration in my voice, maybe you hear that? Sometimes I get an irritation and I need to cough.'[34] Otherwise, she says, she feels no sensation. She swipes the magnet once each morning, then doesn't need to take any medicine for the rest of the day.

Monique is part of a pioneering trial of Tracey's big idea. The research is being run by Paul-Peter Tak, a rheumatologist at the Academic Medical Center at the University of Amsterdam and GlaxoSmithKline. Tak started with a pilot study of eight patients with long-standing rheumatoid arthritis, who had failed all other treatments. Their implants delivered 60-second bursts of vagal nerve stimulation (VNS), once per day for 42 days. Tak reported in 2012 that six of the patients benefited significantly, with improved symptoms, and reduced levels of inflammation in their blood.[35]

Monique is part of a second trial of 20 patients that made news headlines in January 2015. Tak told journalists that 'more than half' of these patients significantly improved, including Monique. Before the trial, even on the best available medication, she struggled to walk across the room. Now, with no drugs at all, she is pain free. 'I have my normal life back,' she told Sky News.[36] 'Within six weeks I felt no pain. The swelling has gone. I go biking, walk the dog and drive my car. It is like magic.'

As I write, these results have not yet been published in a scientific journal, and without a placebo group it's hard to know how much of the patients' improvement was really down to the VNS. Tracey (who is now president of the Feinstein Institute for Medical Research in Manhasset) is optimistic about its potential, however. Human trials are also now underway for Crohn's disease, and Tracey believes that in principle, VNS could work for any disorders that involve damaging inflammation, such as psoriasis, multiple sclerosis – and sepsis and septic shock. Anti-inflammatory drugs don't work in everyone and can have serious side effects, in large part because they suppress the immune system not just where it is needed but throughout the whole body. Stimulating nerves may ultimately allow much more focused treatment, says Tracey, by targeting only those nerve fibres running to particular locations.[37]

In theory, electrical stimulation could be used to modulate other branches of the immune system too – and in fact *any* aspect of physiology that is under nervous system control. Researchers have already found that in an animal model of haemorrhage, VNS triggers production of thrombin (an enzyme involved in blood clotting) at the injury site – suggesting it could help to stem uncontrolled bleeding during surgery or after trauma.[38] Meanwhile delivering electric current to the nerves controlling the gut might help IBS patients,[39] and some researchers have speculated that manipulating nerve signals could delay the progression of some cancers.[40]

VNS also shows some promise for psychiatric conditions. The technique is already widely used to treat epilepsy, and intriguingly, people receiving it tend to report better mood (independent of any effects on their seizures). Tak saw improved mood in his rheumatoid arthritis patients, too. This phenomenon has led to studies assessing whether VNS can help people with treatment-resistant depression.[41] The evidence so far is limited, but trials suggest that it does benefit some patients, although it can take several months to see an improvement.

Tracey calls this new field 'bioelectronics' and claims that we are witnessing a revolution in medicine, in which we move away from treating diseases with chemical drugs and start using electrical signals instead. 'I think this is the industry that will replace the drug industry,' he told the *New York Times* magazine in 2014.[42]

It's a bold idea, but a lot of people seem convinced. Scientists I talk to rave about Tracey's work. Publications from *Forbes* to *Scientific American* have splashed on his story.[43] And while biofeedback struggles for funding, companies and governments are throwing money at the idea of implanted bioelectronic devices. In 2013, GSK announced a $1 million innovation prize (in addition to $50 million it was already spending on research), and the NIH announced a seven-year programme worth $248 million; in 2014, a new DARPA initiative was highlighted by President Obama.[44]

Meanwhile Tracey has started a scientific journal dedicated to bioelectronics, as well as a company, SetPoint, which aims to develop miniature injectable nerve stimulators, perhaps as small as a grain of rice, that will be charged wirelessly and controlled via an iPad. The idea is that eventually these devices will work in real time, to monitor the incoming signals travelling along our nerves and, where needed, modulate their output to our organs.

But what about the conscious mind? Can we learn to harness the inflammatory reflex with our thoughts?

Tracey has argued that in theory, we might. Back in 2005, he suggested that insights from his work could help to direct research into mind–body therapies,[45] and this is now starting to happen. For example, several scientists are studying whether techniques like biofeedback and meditation, which influence vagus nerve activity, can reduce inflammation through this pathway.[46]

For severely injured patients or acute situations such as septic shock, a strong, swift dose of electrical stimulation is probably going to work best. But Tracey suggested that for chronic illness – anything from hypertension to rheumatoid arthritis and inflammatory bowel disease – perhaps we might take a longer-term, preventative approach, using techniques such as meditation or biofeedback to improve vagal tone gradually over time.

I don't know what Tracey thinks now about the potential for psychological approaches – he declined an interview for this book so I wasn't able to ask him. He doesn't mention mind–body therapies in more recent articles, and instead suggests that the tiny injectable devices his company is developing will become routine.

Researching both methods makes sense to me, however. The potential for bioelectronics in medicine sounds truly exciting, but understanding how mind–body techniques can influence the nervous system over time might help in less extreme cases to avoid the need for stimulators – a highly invasive solution, after all, that would leave millions of us dependent on expensive implants with significant medical risks (not to mention security concerns – a 2014 *New York Times* article pointed out that putting our nervous systems under wireless control might leave them open to being hacked).[47]

Either way, though, thanks to Tracey's work, the role of the brain and the nervous system in health is finally moving centre stage. And the potential for transforming the treatment of so many conditions is something that he feels finally gives some meaning to Janice's death. He thinks of her, he said in 2005, as 'like an angel'.[48] In his research, and in the patients he helps, she lives on.

12

LOOKING FOR GOD
The Real Miracle of Lourdes

We wheel her in on a gurney. She's in her nineties, perhaps, with pale, squashy flesh and gnarled hands and feet; her face all thread veins and no teeth. The bed nearly fills the square cubicle. Behind her is the blue-and-white striped curtain she entered through. On either side, the tiled walls are lined with plastic chairs and hooks. Ahead, past her feet, is another curtain.

She's shaking as we undress her, unbuttoning her blouse to reveal a voluminous tummy. 'Ne vous inquiétez pas,' Madame – a squat, Spanish lady – instructs her. Don't worry.

Soon she's naked except for an enormous nappy. Two of us stand on each side, working together, our moves choreographed and rehearsed. We roll her one way and then the other as we slide a sheet underneath her hefty body. We place a blue blanket over her, lift her up on the sheet to slide a stretcher underneath, then we whisk the blanket away and over goes another sheet like a tablecloth, except that this one is cold and wet.

It takes seven of us (three on each side plus one at the head) to carry the stretcher, feet first, past the inner curtain and into a second chamber. It's a small, austere space, lined with grey stone. Square but with a high, curved ceiling, it gives the impression of a miniature chapel.

The floor is tiled, wet and treacherous, and in the middle is a rectangular stone trough, filled to knee height with cold, blue-tinged water. A small, blue-and-white statue stands at the far end: the Virgin Mary. We shuffle down a couple of steps until the woman's stretcher is over the water, head resting on the top step. Then we count together, *un*, *deux*, *trois*, and plunge her into the water.

I've been doing this all day; dipping woman after woman into these icy baths. This little space is the last of a row of ten or so similar curtain-lined cubicles, each with its own team led by a Madame. We're all unpaid volunteers, and it's unlike any job I've done before. We start each shift with 20 minutes or so of singing and prayer, voices sailing up above the cubicle walls.

Then the women come in (there are separate baths for men, and one for children). They've been queuing for hours for this moment, and they've travelled from around the world, just as the volunteers have. They're American, Italian, Indian, Irish. Young, old, well, sick. They're all here in the belief that these waters have healing powers. This is Lourdes.

A small town in the foothills of the Pyrenees mountains in France, Lourdes was relatively unknown until 1858. Then at this remote grotto, a 14-year-old girl called Bernadette claimed to have several visions of the Virgin Mary, and, according to the story, a spring began to flow at the site. Lourdes is now one of the major pilgrimage sites of the Catholic Church. More than five million people come here every year, looking for spiritual – and physical – healing. Three intertwined churches now rise out of the rock around and above the grotto, and there's a series of fountains where visitors can drink the blessed water. But for most, their experience centres on the baths.

Many religions have holy places where devotees travel in the hope of healing, and to wash away their sins. Millions of Muslims congregate in Mecca in Saudi Arabia for the annual Hajj pilgrimage, while Hindus gather every 12 years at the Ganges river in India. Other Catholic destinations sparked by apparitions of the Virgin Mary include Medjugorje in Bosnia and Herzegovina, and Fátima in Portugal. But Lourdes is unusual, perhaps unique, among religious pilgrimage sites, because it claims to validate scientifically any cures that happen here.

If someone claims a dramatic recovery in Lourdes, a committee of physicians collects relevant medical records and investigates whether there is any possible scientific explanation. If not, a bishop then decides whether to afford the unexplained cure the status of a miracle. Since 1858, more than 7,000 people have reported themselves cured to the committee, and 69 of them have been stamped as miracles. These lucky few have apparently been freed from ailments including tuberculosis, blindness, multiple sclerosis and cancer.

I'm interested in these apparent cures. I don't personally believe in miracles, at least not the kind that defy nature's laws. But these cases raise a profound question: can religious experience and belief affect our brains and in turn our bodies? Lourdes seems a good place to look.

I start in the baths. We work three-hour shifts, hot and crowded in the cubicles, as pilgrim after pilgrim appears through the curtain. We ask the women to undress, and wrap a sheet around them. Doing our best to communicate in sign language, we help each with her buttons, undo her shoelaces, unhook her bra. Then, one by one, we usher them through the inner curtain. In an organised, mechanical set of movements we walk them to the far end of the bath and swoosh them backwards. Some women cry, some cry out, as they're plunged into the cold water. Some touch and kiss the Mary statue. Some are stiff and tense, resisting the water; others throw themselves back with such force that we barely catch them before their heads hit the tiled steps.

One American woman stands for a long time, whispering to the statue. A Nigerian mother sobs and asks me to pray for her son. We turn them around, saying prayers as we walk them out of the water. Then we dress them and send them out through the curtain into the spring sun.

**

Does believing in God make you healthier? It's fair to say that this question has not been a top priority for scientists. The word 'spirituality' barely appeared on the PubMed database (which collates the world's biomedical journals) until the 1980s. High-profile scientists such as Richard Dawkins and Stephen Hawking have written entire books dedicated to eradicating the need for God.[1] According to one scholar who works in this area, investigating the relationship between religion and health was seen until recently as an 'anti-tenure' factor,[2] a sure way to send your career crashing.

But in recent years, there has been a surge of interest. Thousands of studies on the topic have now been published in major medical and psychiatric journals, while US medical schools routinely offer courses on religion, spirituality and health.

Much of this research concludes that being religious leads to better emotional or psychological health. But an increasing number of studies are claiming physical benefits too. In the last few years, believing in God has been linked with lower rates of heart disease, stroke, blood pressure and metabolic disorders, better immune functioning, improved outcomes for infections such as HIV and meningitis, and lower risk of developing cancer. Religious people have lower risk of cognitive impairment and disability with age, faster recovery following surgery, and lower rates of medical service use.[3]

Based on these results, some scholars argue that religion should be integrated into the medical system, with doctors enquiring about and supporting their patients' spiritual health. But critics like Richard Sloan, professor of behavioural medicine at Columbia University in New York, and author of *Blind Faith: The unholy alliance of religion and medicine*, argue that many of these trials don't adequately tease out other factors that aren't directly connected with belief in God.[4] Religious people tend to have healthier lifestyles, for example – they drink less, smoke less and have less unprotected sex.

What's more, big surveys often use church attendance as a measure of how religious someone is. In general, religious attendance is associated with 7–14 years of additional life.[5] But you need a certain level of health to get to church in the first place, so perhaps it's not surprising that this group lives longer. Those who attend church may also have stronger social bonds, and as Sloan points out, 'There are plenty of other ways to enhance social connectedness.'[6]

On the other hand, a recent meta-analysis of 91 studies tentatively concluded that even after accounting for these factors, 'religiosity/spirituality' may have a protective effect in initially healthy people, with those who attend church regularly around 20% less likely to die (when followed for periods of five years or more) than those who don't.[7]

If there is an effect, it might be partly due to placebo responses; improvements in health triggered by the belief that God will heal us. A 2011 poll of more than 900 American adults found that 77% of them believe prayer can help to heal people from an injury or illness.[8] Belief in fake treatments banished the symptoms of Linda Buonanno's irritable bowel syndrome and Bonnie Anderson's fractured spine. Similar biological pathways presumably help many of those who pray, or visit a pilgrimage site such as Lourdes.

But I soon find there's a lot more to it than that.

***

Sheri Kaplan describes herself as 'a nice Jewish girl'. She's pretty with blue eyes and curly red hair. She grew up in Florida but spent her mid-twenties in Manhattan: partying, dating and working for a magazine. After that she returned to Miami, started a catering business with her sister, and settled down with a steady boyfriend. Then in 1994, when she was 29, everything changed. She was diagnosed with HIV.

'I was numb,' she said in an interview in 2005. 'It's like being hit by a train – there's confusion, fear, anger, grief, sadness.'[9] Her boyfriend left. She was convinced she was going to die. She gave up the catering business and maxed out her credit cards on a two-month trip to Europe. She thought it was a final fling, but it turned out to be a new beginning, and she returned to Miami determined to make the most of whatever time she had left. She looked for a patient support group but couldn't find one that catered for heterosexuals with HIV – they were all aimed towards gay men or drug addicts. So she founded her own.

The Center for Positive Connections started as a handful of women Sheri found through local clinics, who met each week to chat over coffee. A few years later, the group had a half-million dollar budget and over 1,500 members. It provided social activities, support groups, national hotlines, personals ads and an annual Caribbean cruise. Sheri travelled the world with her work, winning awards and meeting celebrities like Richard Gere.

And through the group, she found a new purpose in life, interpreting her illness as part of God's plan. 'I got HIV because it is my purpose of being,' she said. 'I had to understand what it is like so I can help the community on a different level and help create social change.'[10] Amazingly, she stayed well, and she believed her religious faith was helping to keep her virus at bay. She wasn't alone – a 2006 study found that 50% of HIV patients thought their religion/spirituality was helping them to live longer.[11] But was she right?

It sounds a bit crazy. HIV infects the immune system's CD4 cells, using them to make thousands of copies of itself and killing them in the process. Eventually the number of CD4 cells in the body drops so low that the immune system stops working, leaving patients vulnerable to life-threatening illnesses. Treatments available today allow many people infected with HIV to live a long and healthy life, but in the mid-1990s before those drugs were available, infection was generally seen as a death sentence.

But Gail Ironson, a psychologist at the University of Miami in Florida, noticed that some of the patients she saw didn't get sick. Many of those patients talked about the importance of spirituality in their lives, and she began to wonder whether it really did influence their health.

Ironson interviewed around 100 recently diagnosed HIV patients, including Sheri, about their lives and beliefs, then followed their progress for four years. She found that 45% of the patients became more religious after their diagnosis, while 42% didn't change their beliefs significantly and 13% became less religious. Ironson's hunch turned out to be right. Those patients who became more religious lost CD4 cells much more slowly over the four years, and had lower counts of virus in their blood.[12] Take Sheri, for example. In 2005, 11 years after she was diagnosed, she still had no symptoms, and enough CD4 cells that she did not need to start HIV medication.

Changes in religious belief are likely to alter behavioural factors that could in turn influence disease progression, such as living healthily or taking regular medication. But Ironson says that her result was significant even after accounting for differences in lifestyle, medication and other psychological factors such as optimism and depression.

This study isn't conclusive on its own and, as far as I know, nobody has tried to replicate Ironson's result. If she's right, though, there's no need to invoke divine intervention to explain why patients who turned to God did better. Ironson believes instead that this lowered their levels of stress.

There's substantial evidence that stress accelerates the rate at which asymptomatic infection with HIV progresses to full-blown

AIDS. In particular, the stress hormone noradrenaline helps the virus to enter CD4 cells, and to replicate faster once it is inside.[13] In one high-profile study,[14] which followed HIV-positive men for nine years, each extra (moderately severe) stressful event increased their risk of progressing to AIDS during that time by 50%. Some trials suggest that reducing stress levels through meditation or cognitive behavioural therapy can slow the progression of the disease.[15] Trusting in God may work through the same pathway.

In fact, the apparent health benefits of religion – including reduced risk of chronic diseases such as diabetes, dementia and stroke – look very similar to those you get from lowered stress. Neuroscientist Andrew Newberg of Thomas Jefferson University and Hospital in Philadelphia, who studies the effects of religion on the brain, tells me that like meditation, prayer lowers heart rate and blood pressure, and helps us to regulate our emotional responses to stressful situations. Religion helps believers 'understand themselves, it helps them to understand the world, it helps them to cope with things,' he says.[16]

Believing in God may also provide us with the ultimate social support in the face of adversity. 'There's a sense of someone else beyond you who is your fall guy,' says Michael Moran, a Catholic doctor from Belfast who is a member of the International Medical Committee of Lourdes and volunteers regularly. 'At times, it almost feels like you're being taken into someone's arms and held.'

But, warns Newberg, as with the placebo effect, religious belief has a dark side, for example if you're in a church or religious group that espouses hatred and anger towards others. 'Those are typically very negative emotions which can be detrimental to the person's brain and body.' Reducing stress and benefiting health, Newberg argues, requires, 'being religious in a way that espouses positive emotions; emotions of love, compassion, connectedness, a sense of unity and so forth with other people. Not only people in your group but people outside of that group.'

Even within mainstream religions, going through a spiritual

struggle or believing in an angry or judgmental God seems to make people more stressed, with subsequent effects on their health. In a 2001 study, psychologist Kenneth Pargament of Bowling Green University in Ohio followed nearly 600 hospital patients aged 55 and over for two years.[17] Those who experienced spiritual struggles related to their illness – wondering whether God had abandoned them, questioning God's love for them, or deciding that the devil had made them ill – were more likely to die in that time, even after controlling for other factors.

Meanwhile Gail Ironson asked the HIV patients she followed how they viewed their God. (She didn't explore the atheists in the sample because there were so few of them: 6.3%.) She measured their responses on two separate scales: whether they saw God as 'positive' (benevolent, forgiving and merciful) or 'negative' (a harsh judge who would punish them for their sins). Those who viewed God positively – like Sheri – had significantly slower disease progression, with five times better preservation of CD4 cells, than those who did not.[18] By contrast, those who viewed God as harsh and punishing lost CD4 cells more than twice as fast as those who didn't. The effects were significant after taking into account other aspects of lifestyle, health and mood – in fact the patient's view of God predicted disease progression better than any other psychological factor that Ironson measured.

One of the participants who felt abandoned by God was Carlos, a man with a Catholic background who had just moved to New York to study for a BA degree when he was diagnosed with HIV. 'I had no friends in New York so I had to deal with it on my own,' he told Ironson. 'Any belief that I had in a higher being or in a spiritual presence was completely extinguished . . . I felt like I was being punished. I thought I was going to die for my sins.'[19] Unlike Sheri, who remained symptom-free for the duration of the study, Carlos's infection progressed rapidly to AIDS after he was diagnosed.

***

It's just before dusk and I'm standing across the river from the grotto at Lourdes. I'm in front of the Accueil Notre Dame, one of the hospitals here that caters for sick pilgrims, waiting for the Procession of the Blessed Sacrament to begin. A group of priests appears, reverently carrying the round, white wafer in its gold stand, sheltered beneath an ornate gold-and-cream canopy.

It's busy. The priests congregate on an open area of ground in front of the hospital, joined by a group of people in wheelchairs and on stretchers. A crowd of other pilgrims and tourists forms a circle around them, while many more sit on the wall in front of the river, in a line that stretches away down its banks. There's a short service, then everyone starts walking towards the nearest bridge. The wafer goes first, followed by the stretchers – carrying patients with oxygen masks and drips – then the wheelchairs, pushed by nurses in white headdresses and black capes with a distinctive red cross.

After the most sick, other pilgrims fall into line, pulled along in blue, wheeled 'chariots'. There's a girl, maybe 12, dressed in a green anorak and pink jeans, with her hair in a ponytail. She's hunched over, rocking violently back and forth, but holding her mother's hand triumphantly in the air. Just behind her is a little boy, aged two or three, with a mop of blond curls, sucking at the sleeve of his blue cardigan.

Those who can walk follow behind. There are hundreds of us, in a procession that moves slowly across the bridge then snakes around in front of the big, spired church that marks the site of the grotto. Choir music is pumped through loudspeakers along the route. A robust old woman next to me belts out 'Amen! Amen! Alleluia!', while using her umbrella as a walking stick.

Instead of going into the church, we wind down into what looks like a concrete underpass, and I wonder where on earth we're going. Once underground, however, we turn a corner and the passageway opens out into an enormous subterranean basilica, as big as a football pitch (I read later that it can hold 20,000 people). It's made of concrete, with lines of square floodlights and great diagonal rafters.

Scarlet banners hang along each wall, decorated with pictures of the saints. Hundreds of wooden benches line up in neat rows facing a raised central platform, with steps on all four sides that make it look a bit like a pyramid. It's bright in the spotlights and supports a large, white altar; a silver sculpture of Christ on the cross; and a golden ball full of incense, with grey smoke that curls up towards the roof.

Directly in front of the platform are rows of those blue chariots – the sick in pride of place. And now I see the choir, over to one side, accompanied by trumpets and a pipe organ. As the service begins, screens suspended from the ceiling zoom in on the priests around the altar, showing us the action up close; pictures that are also transmitted to the faithful around the world on Lourdes TV.

We sing and chant in a range of languages – Latin, French, German, Spanish – guided by subtitles on the screens. There's lots of standing up, sitting down, joining in. At one point, priests in cream robes walk around with the wafer, holding it up in front of each group of the congregation as they ring a bell. When they get to our section of the hall, the people around me all kneel and cross themselves.

I feel out of place amidst all this singing and signing. I've never attended a Catholic mass, and I usually try my hardest to avoid religious ceremonies. I get uneasy about the idea of substituting reason and clear thinking for robes, incantations and mysterious higher powers. But at the same time, it's beautiful; a hugely impressive assault on the senses. The banks of lights, the colours of gold, red, cream, silver. The sweet but smoky smell of incense; the uplifting music; the enormous crowd. The synchronised physical exertion of standing up and sitting down.

Quite unexpectedly I feel a powerful sense of connectedness, as if I'm at the centre of something much, much bigger. I get the feeling that, in this great hall, we're bound by threads that stretch far around the world as well as forwards and backwards through time. Around me, thousands of people who have never met before today are speaking and singing in different languages yet in perfect timing and harmony. And their pictures are being beamed around the planet so that this moment can be shared by millions more. These chants and movements form a ritual that people have been participating in for centuries, and that will likely live on for centuries to come.

Neuroscientist Andrew Newberg argues that rituals like this are a hugely important component of how religion and spirituality affect us physically as well as mentally. They have such a powerful effect, he argues, because their roots stretch back deep into our evolutionary history. In the animal world, they started out as mating rites. But as our brains have become more complex, he says, we have adopted rituals for other purposes too, from a baby shower to the Olympic opening ceremony. 'Part of what rituals ultimately do is connect us with each other,' says Newberg. Whereas mating rituals connect two individuals, in religious or other cultural contexts they help to bring a society or community together in a common set of actions and beliefs.

When it comes to religion, rituals bind us together so forcefully because they make the abstract beliefs we share seem more concrete. 'When you hold a particular belief you can feel pretty strongly about it,' he says. 'But if it is incorporated into a ritual, it makes it a far more powerful kind of experience, because it is something that you not only think about in your brain but feel in your body.'

This can be as simple as reciting the Rosary, which links a set of religious beliefs with the physical act of counting beads between your fingers. But rituals are probably more powerful when they involve groups of people all doing the same thing together, as we are in this giant underground hall.

Lourdes hasn't turned me into a believer. But after attending this giant underground service, I'm struck by the physical force that religious belief can have. Here in this basilica, a shared vision is translated into something that we can all see, hear, feel, smell (and for those who take communion, taste). Religious belief might be intangible but this ritual has made it a solid thing in the world. Suddenly I don't find it so hard to accept that such belief might have a potent effect on the body, too.

***

If the physiological effects of religious belief can be explained by mechanisms such as stress and rituals, does God even need to be in the picture? We've already seen some of the physical benefits of secularised meditation programmes such as CBCT and MBSR, but do they lose anything in translation?

Hardly anyone has studied this, but psychologist Kenneth Pargament believes the spiritual aspect does make a difference. He and colleague Amy Wachholtz asked volunteers to meditate on a particular phrase. One group of volunteers chose between spiritual phrases, such as 'God is peace' or 'God is love', while those in another group were asked to select a non-religious phrase, such as 'grass is green' or 'I am happy'. The volunteers meditated 20 minutes a day for two weeks, then Pargament and Wachholtz tested their pain tolerance. Those in the spiritual meditation group were able to keep their hands in a bath of ice-cold water for almost twice as long (92 seconds) as either the secular medi-tation group, or people who spent the same amount of time learning a relaxation technique.[20]

In a second study of 83 migraine sufferers, those who practised spiritual meditation for a month had fewer headaches and greater pain tolerance than the secular meditation and relaxation groups.[21] (They also felt less anxious and generally happier.) 'Content counts,' Pargament tells me. 'The spiritual phrase seems to magnify the impact of meditation.'[22] These are small studies that need repeating but if the results hold up, Pargament thinks that a spiritual perspective may help to reduce the emotional impact of the pain by placing it in a larger, more benevolent context. 'It shifts the mind away from physical and mundane concerns to a focus on the larger universe and the individual's place within it,' he says.[23]

Encouragingly for me, Pargament argues that spirituality doesn't have to mean belief in a distinct creator, and that you don't need to be religious in order to benefit from its effects. In his studies, volunteers who didn't want to meditate on a religious mantra could choose an alternative, for example substituting God for 'Mother Earth' (although only one person actually did this). Anything we perceive as having divine character and significance – as set apart or special – should work. In the United States, this is typically understood as meaning some kind of divine figure: God, Jesus, something transcendent, he says. But it can be something else.

In Sweden for instance, nature is often seen as sacred, with people responding to and experiencing nature similarly to how a religious person might experience God through prayer. 'People write about their experience of being outdoors, being one with nature, feeling a pulse in nature that never ends,' says Pargament.

Someone might hold their work as sacred; the idea of a more just, loving world; or their family. Pargament quotes a mother of two young children: 'To see my kids is to realize that they are – well, godlike . . . not because they are particularly unusual children, but because I could not with my own two hands have created anything as wonderful or amazing as they are . . . Just tickling their feet and hearing them giggle – that's cosmic, that's divine.'[24]

Pargament's ideas dovetail with other research suggesting that seeing ourselves as part of something bigger, or having a meaning or purpose beyond ourselves, helps us to do better physically. For example, we heard in Chapter 9 how in a study of a three-month meditation retreat high in the Colorado mountains, stress researchers Elizabeth Blackburn and Elissa Epel found that meditators had higher levels of the enzyme telomerase – which slows cellular ageing by protecting telomeres – than a group of controls. When the researchers looked at which psychological changes might be contributing to this effect, they found that the effect on telomerase was stronger in people who reported feeling a greater sense of control and an increased sense of purpose in life.[25]

The lead researcher of that study, neuroscientist Clifford Saron of the University of California, Davis, argues that this psychological shift towards purpose and control may have been more important than the meditation itself. The participants were already keen meditators, he points out, so the study gave them three months to do something they loved.[26] Simply spending time doing what's important to you, whether gardening or volunteer work, might similarly benefit health. What the study really shows, says Saron, is 'the profound impact of having the opportunity to live your life in a way that you find meaningful',[27] whether that involves God or not.

Meanwhile UCLA's Steve Cole, whose work on loneliness and gene expression we learned about in Chapter 10, has also investigated happiness. He found in one study that people who score high in eudaimonic wellbeing (the satisfaction that comes from engaging in activities with a greater meaning or purpose) have lower expression of genes related to inflammation than those driven by shallower pleasures like shopping or having sex.[28] Cole argues that having a higher purpose may make us less stressed about threats to our personal wellbeing. If we die, the things we care about will live on.

In other words, feeling part of something bigger may help us not only to deal with life's daily hassles but to defuse our deepest source of angst: knowledge of our own mortality. John Cacioppo argues in his 2008 book, Loneliness, that we have an innate biological need for this connection. 'Just as finding social connection is good for us, finding that transcendent something appears to be very good for us, whether it is a belief in a deity or a belief in the community of science,' he says. 'It is only through some ultimate sense of connection that we can face our own mortality without despair.'[29]

Western society tends to value maximising control and breaking through limits, says Pargament. 'We try to solve problems. We try to enhance longevity.' But at some point, we all face events and experiences that we cannot control. And while western medicine has brought huge advances in health and life expectancy, it isn't very good at helping us to deal with those barriers when they hit. 'The classic description is physicians when they realise there is nothing more they can do for their patients,' says Pargament. 'Unfortunately at times they will just walk away. Sometimes even angrily because they face the limits of their control and they don't know what to do with that.'

Spirituality, he argues, fills that gap by helping us to accept that we are frail, finite people. No matter how good medicine gets, 'We're all going to face problems that are somewhat intractable, including physical pain,' he says. 'And eventually, we are all going to die.'

***

I want to know about the miracles.

'If there's one place in the western world where there is a link between science, religion and health, it is Lourdes.' Alessandro de Franciscis, head of Lourdes Medical Bureau, is lounging with his legs crossed and his right arm hooked over the back of his chair. His office, a stone's throw from the entrance to the underground basilica, is spacious and elegant, with sofa and armchairs – coffee-coloured, padded, carved from walnut – arranged around a Persian rug. Behind them, a huge wooden desk sports an upright crucifix and an old-fashioned green desk lamp. On the bookshelf, Richard

Dawkins' *The God Delusion* and Stephen Hawking's *The Grand Design* are prominently placed beneath four framed photos of de Franciscis meeting the Pope.

De Franciscis himself looks scholarly, with a high forehead and dark, greying hair. A former paediatrician from Naples, Italy, he also has a degree in epidemiology from Harvard. He's charming but combative – he repeatedly fires names of scholars at me to see whether I'm familiar with their work, and refuses to be interrupted as he tells long stories, like a meandering but unstoppable train.

If anyone reports a cure in Lourdes, de Franciscis oversees the process of checking it. The bureau was created in 1883, 'as a defence against the accusation of Lourdes as a place where there was too much superstition, too many miracles,' he says. 'In France they felt very proud of being the country in which they had invented modern times. Rationalism, Cartesianism, all of that. That there was a place where people came in growing numbers to experience prayer and in some instances cures, it was disturbing.'[30]

Doctors affiliated with Lourdes began to examine and document every cure that took place. The aim, says de Franciscis, was nothing less than 'proving the existence of God through the power of scientific explanation'. Pope Pius X sanctioned the endeavour in 1905, when he decreed that claims of cures at Lourdes should be submitted to 'a proper process'.

Led by de Franciscis, that process continues today. If a cure is reported, he convenes a meeting of affiliated doctors present in Lourdes at the time. They gather information on the case. Was the person definitely sick? Are they definitely better? Did anyone witness the moment of the cure? They also request records and scans from the claimant's home country, from before and after their Lourdes visit; check whether the person has received any treatment that might explain their recovery; gather medical opinions; and wait, sometimes decades, to ensure that the cure is lasting.

If the bureau doctors are satisfied, they send their notes to the International Medical Committee of Lourdes (CMIL), which votes on whether the event is definitely an unexplained cure, and produces a formal medical report. As doctors, that's as far as they can go, says de Franciscis. The report is sent to the claimant's local bishop to decide whether the cure represents a divine miracle.

De Franciscis argues that this process gives Lourdes 'a seriousness, an accountability in the relationship between faith and medicine' that isn't found at any other pilgrimage site in the world. I had wondered if he might be defensive about the medical evidence supporting the 69 miracles announced so far, but he's clearly proud of the reports and gives me copies of anything I want to see.

The documents he hands me include reports on cures from conditions such as cancer, blindness and paralysis dating from throughout the twentieth century. It strikes me that several of the original diagnoses, including that of a Frenchman said to have multiple sclerosis, are based on the patients' experience and symptoms rather than physical tests. These people were dismissed by their doctors as hopeless cases, and the cures were clearly miraculous to those concerned. But rather than providing proof of paranormal or divine intervention, I wonder if what these recoveries really demonstrate is the power of the mind to impose devastating symptoms of illness upon us, and the power of religious belief to lift that burden.

One of the reports I look at is different. This is the case of a young Italian soldier called Vittorio Micheli; the sixty-third miracle, recognised in 1976. Vittorio was admitted to hospital in April 1962, when he was 22, with pain in his hip. He was diagnosed with a malignant tumour – an osteosarcoma – in his pelvis. Over the next few months, the tumour destroyed the bones in his left hip, and invaded the surrounding muscle. According to the report, Vittorio received no treatment for his cancer – no surgery, chemotherapy or radiotherapy. With his leg dangling precariously from his body, doctors encased him in a plaster cast from head to foot, and said there was nothing they could do.

At his mother's insistence, Vittorio went to Lourdes in May 1963. He was frail and desperately ill – not eating, on high doses of painkillers, confined to his stretcher. 'When I was there in Lourdes, I didn't feel anything extraordinary,' he recalled in 2014. 'But on the way back, on the train, I didn't need any painkillers any more. And I started to be hungry. I started to eat again.'[31] He returned to hospital in Italy, but his doctors didn't pay much attention to his story until months later, when Vittorio started to feel that his leg was once more attached to his body. In February 1964, his doctors removed the cast, and he was able to walk. Vittorio has returned to Lourdes many times since, often as a volunteer, and even got married there. Now in his seventies, he can still walk perfectly well.

This is the case I choose to investigate further. Once I'm back in the UK, I show Vittorio's report to Tim Briggs, an orthopaedic surgeon and expert on osteosarcoma at the Royal National Orthopaedic Hospital in Stanmore, Middlesex. Radiographs taken before Vittorio's visit to Lourdes show an enormous mass of tumour covering his left hip, with the head of the femur and the socket of his pelvis completely eaten away. Histology samples – slices of cells from the surrounding tissue – show rampant, invasive cancer. Then, over the page, an X-ray image taken after Vittorio returned from Lourdes and his cast was removed. There's no longer any sign of cancer, and the bone has regrown. The bone is a bit misshapen, like scar tissue, but the structure is there – the ball of the femur, the socket of the pelvis – all perfectly functional.

Briggs seems impressed at first. 'It's astounding,' he says. He takes the report to study in detail, and a few weeks later calls me back to his office. 'I've got an answer for you!' he says, triumphantly. After studying the histology slides, he confirms that the cancer was indeed malignant, but says that rather than osteosarcoma it looks like lymphoma, a more common cancer that affects not the bone cells themselves but lymphocytes, a type of white blood cell, within the bone marrow.

Lymphoma is 'a completely different malignancy', says Briggs. Osteosarcoma is aggressive. Today it is treated with chemotherapy, surgery to completely remove the tumour, then more chemotherapy, and even then only 60% of patients survive for five years. If Micheli's tumour had been osteosarcoma, 'He would have been dead.'[32]

By contrast, patients with lymphoma don't usually require surgery, and it can respond very well to chemotherapy. What's more, Briggs and his colleagues have spotted a mention, buried deep in the Lourdes report, that Micheli may have received a drug called endoxan. This is another name for cyclophosphamide, an immunosuppressant drug often used to treat lymphomas because it kills white blood cells. The report is ambiguous on this point – elsewhere it states that Micheli was not treated for his cancer – but for Briggs, the only plausible explanation is that the mention of endoxan is accurate. Micheli 'obviously had a very good response' to it, he says.

Once the cancer was gone, Briggs isn't surprised that Micheli's hip joint regrew. 'After chemotherapy, bones are amazing in their capacity to regenerate,' he says. He reckons that in Micheli's case this might have taken around 6–12 months. From the limited information in the medical report, it's impossible to know for sure what happened. But although Micheli's recovery must have seemed miraculous to him at the time, there does not seem to be anything here that science is categorically unable to explain.

Back in Lourdes, De Franciscis insists that even if we subsequently find medical explanations for some of the miracles, it won't change their status as signs of divine intervention for those who want to believe. 'A miracle is an interpretation,' he says. 'A bishop believes that the person has received a gift from God.'

I'm left feeling rather confused about the point of scientifically validating cures in the first place. For all of the committee's meticulous work, it seems to me that the existence of divine miracles remains a matter of faith rather than science. What de Franciscis and I do agree on, however, is that religion – with Lourdes as a shining example – is a powerful mix of all of the ways in which the mind can benefit health, including social connection, stress reduction and placebo effects. And while this may well help people to feel better in all sorts of ways, it also proves that the mind cannot, in general, produce miraculous cures. Lourdes pilgrims, for all their belief, don't generally go home physically transformed. Of the hundreds of millions of pilgrims who have visited, after all, there have been just a few thousand reported cures, and only 69 claimed miracles.

'If Lourdes regarded itself as being a clinic, it should have closed the day after, it is a complete failure!' says de Franciscis. 'No, Lourdes is not a clinic. It is a place of worship.' Counting cures misses the point, he argues; there is something much bigger, more transformative, that Lourdes has to offer. The original purpose of the medical bureau, he says, was to document miracles and therefore prove the existence of God. But now de Franciscis has a different mission.

***

I say, 'Hello,' and Christopher beams at me. He is 24, but looks much younger. He's small, crunched up in his wheelchair with a hunchback and frail limbs. He pulls at my hand, then points at his huge smile to encourage me to take his photo on my phone. I show it to him and he holds the screen just an inch or two from his crossed eyes to scrutinize the picture before nodding, he's pleased with the result.

Christopher was born with a rare genetic condition called Rubinstein-Taybi syndrome. A mutation in a single key gene causes wide-ranging problems including mental retardation, stunted growth, heart problems, and difficulties with breathing, feeding, vision and speech. Christopher can't walk or speak, his mum, Rose, tells me. He's in nappies, and he needs constant care.

I'm still taking this in as Rose introduces me to her daughter, Mary-Rose, who is three years younger than Christopher. She too was born with a devastating genetic syndrome, unrelated to her brother's condition. Her body is riddled with benign tumours that damage the organs they inhabit, from her eyes and brain to her heart and lungs. Mary-Rose is taller and heavier than Christopher. She's wearing a pink tracksuit, and her blonde hair is adorned with pink and orange flowers. Like Christopher, Mary-Rose uses a wheelchair, wears nappies, and can't speak. She can't feed herself. She has epilepsy, and she's blind. I hold her hand and tell her that I love the flowers in her hair, but her blank expression doesn't change.

Rose and her children are from County Cork in Ireland. They're in Lourdes as part of a 130-person pilgrimage run by a small Irish charity called Casa. We meet in the hotel lobby just after dinner, towards the beginning of their week-long stay. Rose is bright and down to earth but her dark eyes scream of exhaustion.

She first brought Christopher here when he was four months old, she tells me. He had never left the hospital, and the doctors said he had just one month to live. Accompanied by a medical team, she took him to the airport and boarded a plane to Lourdes. 'It's the nearest place to heaven,' she says. 'I wanted him to have no pain when he died.' Now she brings her children every year. In Lourdes, Rose says, they can be accepted. 'People at home don't understand that they're loving individuals. They just see the wheelchairs. Here, people take them into their hearts and love them.'

Since that first visit, Christopher has defied expectations. He has been through 17 rounds of surgery, says Rose, 'on his heart, lungs, legs, ears, eyes – all parts of his body'. But she attributes much of his progress to their annual pilgrimage. 'Christopher wouldn't be alive other than coming to Lourdes,' she insists. 'He was never supposed to eat or interact. Now look at him.' Her son is indeed having a ball, spinning around the lobby in his wheelchair, the darling of the other guests. Meanwhile Mary-Rose 'was getting 40 seizures a day', says Rose. 'Now she is getting three. Her first smile was at the grotto when she was nine months old. Then I knew she was going to be ok.'

It seems to me, though, that the one who needs Lourdes the most is Rose herself. She looks after both of her children full-time at home, as well as her husband, who she says is severely ill. 'At home, I can't go anywhere,' she says. 'I can't push two wheelchairs.' This pilgrimage is the only break she gets. 'I have no other life,' she says bluntly. 'Here, I can be Rose. I'm a person as well. Once the children were born, people at home forgot I existed. When I'm here, I know I exist. Our Lady knows I exist.' Without this yearly trip, and the support of the medical staff and other pilgrims, she doesn't know if she would have the strength to carry on. 'There's a big weight on my shoulders, and by the time I get home, it is gone,' she says. 'I don't know if it's the baths. The grotto. The hugs. The looks. But it's here it happens. I go home a new person.'

The striking thing about Lourdes, I'm finding, is that cured or not, everyone I talk to here feels that they have experienced a miracle.

Over the river in the Acceuil Notre Dame, another Irish pilgrimage is packing to leave. Here I meet Caroline Dempsey from Dungarven, a 47-year-old teacher with short, fair hair and purple crocs. She's sharing a hospital room with three women in their eighties. We sit on her bed and chat as the nurses bring round biscuits and mugs of tea.

Her partner died of cancer. And now Caroline has a sarcoma. It appeared seven years ago in her leg and was surgically removed, but it's back in her abdomen. Caroline isn't deeply religious; she didn't want to come to Lourdes and is here only because her mother insisted. But now she doesn't want to go home.

'There's something different about the Masses here,' she says.

'At home, it feels like it's just a ritual, nobody really means it. But here they are genuine. It feels as though thousands of people are praying for you to be well.' She didn't get much out of the baths. But earlier that day, she had attended an anointing Mass, and as soon as the priest approached her with the oil, 'I couldn't move. I was melting. I felt a huge release.' When Caroline's cancer returned, 'I was so afraid,' she says. 'But today I got this feeling. Have hope, live life, don't dread it.'

Across the corridor is 82-year-old John Flynn. He's bald, and breathless as he talks. Scattered on his bed is a dizzying array of pills and capsules – pink, white, red-and-green; packaged in yellow bottles, foil sachets and white plastic jars. John worked in an iron foundry for 30 years until he tore the tendons in his shoulder and could no longer work. That's when he first came to Lourdes, in 1988. 'I loved the experience,' he says. 'I got hooked.' Now he's back for the sixteenth time.

He suffers from nerve pain, a dead hand and dragging leg after a stroke seven years ago, and arthritis everywhere. At home, he says he gets frustrated about the things that he can no longer do. But coming to Lourdes puts things into perspective – you see people far worse off than you, he says – and helps him to accept the situation that he's in.

Outside the hospital, women sit in the darkness, smoking and chatting as they look across the river towards the floodlit grotto. I'm ushered to a spare wheelchair by Joan – 'I have MS and cancer and diabetes and arthritis,' she says. 'I'm 56.' – and Ann, who suffers from recurrent depression. 'My brother drowned when I was four,' she says. 'My father died when I was seven. I was sexually abused as a child. I got married, then my husband went off with someone else.'

For them, what's special about Lourdes is the social support. The chance to talk through problems is something they don't feel is offered by the medical profession – or by society – back home. The two women only met this week, but 'We've shared our lives,' says Joan. 'When I'm at home, I'm on a journey alone. Here, there's a great togetherness.'

'At home, no one talks to each other,' echoes Ann, with a mug of tea in one hand and a cigarette in the other. 'If we go to the psychiatrist, we're given a pill. The psychiatrist said to me, "I'm not here to listen. I'm here to diagnose and prescribe."' She's not criticising the medical profession, she says. Doctors can't listen to every story, and without psychiatric hospitals she wouldn't be here today. 'But here, there's less fear. People are not afraid to talk. Love is oozing out of the walls.' What every single one of the pilgrims also mentions is the care and support they get from the volunteers here, from teenaged helpers to senior doctors. 'It is fantastic,' says Joan. 'They treat you with respect.'

Surprisingly, I hear similar sentiments from the volunteers themselves. They pay their own way and give up a precious week of holiday to come here every summer. 'I don't come here for the pilgrims,' one volunteer tells me as we line up for dinner in the staff cafeteria. 'I come here for me. Because I need this in my life every year.' Another volunteer, a London banker who doesn't tell his friends that he comes to Lourdes, says he first came as a teenager, to give thanks after he recovered from an illness. Decades later he still comes, because he gets 'such a buzz out of helping'.

Volunteers and medical staff tell me that Lourdes puts their own life concerns into perspective, and provides a camaraderie they don't get in their normal lives. It is the kind of place where you make best friends immediately, they say. Everyone here – sick and well – is equal, no matter what they do back home.

I've seen what they mean. Healthy, sick, rich and poor really do mix intimately here in a way that I've never experienced, and random acts of kindness are the norm. In the baths, volunteers tie pilgrims' shoelaces. In the basilica, the sick are lined up right at the front. Even in the back streets full of tacky tourist shops, there are wheelchair lanes instead of bicycle lanes. A nun I've never met before secretly pays for my lunch. I go to the station to help pilgrims off the train, and find out later that my fellow volunteers included a CEO and a dustman.

This, says de Franciscis, is the real miracle of Lourdes.

In western society, he argues, the sick are sidelined, and stripped of their humanity. 'Once you get admitted to a hospital,' he says, 'You become a leukaemia. You become a hypercholesterolaemia. You become a diagnosis.' Whereas in Lourdes, he reckons, the sick are treated not as diseases but as people, equal to the most senior doctor. 'It is normal in Lourdes to sing together, to pray together, to chat, to dance, to have beer.'

This, then, is de Franciscis' new mission. As head of Lourdes' medical bureau he still documents unexplained cures. But his priority now is to show the wider world the benefits of an approach in which the sick are respected, valued and cared for by all. He ultimately hopes to transform how the sick are treated not just in hospitals and clinics but also in everyday life; to inspire a different way for us all to live that doesn't necessarily involve religious belief. 'It's above the church,' he says. 'This is a different model of society.'

It's a model in which our biological state is entwined with psychological, emotional and spiritual health. This seems to add up to a different kind of healing, one that goes beyond cells and molecules to encompass our humanity too. We've seen some examples of that throughout this book, with researchers repeatedly finding that people treated in a more holistic way do better physically as well as emotionally. Here in Lourdes, this approach is being played out on a grand scale. And millions of patients, volunteers and medical staff alike are coming back year after year just to get a taste of it.

***

It's hot in the baths, and I'm dripping with sweat as I near the end of my shift. This afternoon we've lifted stretcher after stretcher.

It is hard work, physically and mentally. Trying to understand the instructions barked in French. Trying not to slip on the wet tiles. Navigating a whirlwind of underwear of all shapes, sizes and designs. And now here is the old lady with the mountainous tummy.

Her eyes widen as we plunge her into the water. 'Ohhhhh!', she says, her toothless mouth forming a perfect circle. She's only in for a second, then we lift her out and tilt up her stretcher. She fixes her eyes on the Mary statue as the water rolls off. We all speak together: 'Notre-Dame de Lourdes, priez pour nous! Sainte Bernadette, priez pour nous!' Then we remove the wet sheet that covers her and replace it with the blanket.

As we carry her back to the gurney she is calm, no longer shaking, and she grips my arm tightly as the others dress her. 'Merci!', she says. She pulls me close and smiles. 'Merci!' Her eyes are pale grey. Before I saw only the ugliness of age: wrinkles, fat, wasting muscles, dwindling limbs. Now I see kindness, love, laughter, and I'm struck by her beauty. I wonder who she is, what she has done in her life, who she has known. What it's like to be so close to death.

I'm not sure what to say. I have limited knowledge of French, and of her faith. 'C'était parfait,' I whisper. It was perfect.

CONCLUSION

'Can you see it?' Mary Lee McRoberts arranges herself against the wall. 'Drop your lids and dull your eyes,' she advises. 'You don't want to look hard.' We're in a small, darkened room in McRoberts' home in Mill Creek, an exclusive community in Washington state. The space is lined with bookshelves and dominated by a tall massage table on which I'm lying, padded with cushions and covered with a soft, velvet blanket. McRoberts is a reiki master, and she's trying to show me her aura.

The local TV news recently featured McRoberts apparently curing a patient with fibromyalgia.[1] The report described how she works with people's energy fields to clear blockages and heal the body. Her patient, a blonde executive named Sue, says her pain subsided after just a couple of sessions with McRoberts. Sue also lost weight after the reiki therapy, and says blood tests show she has improved on cholesterol and blood sugar too.

I have to admit I'm sceptical about the existence of auras and healing energy fields. There's no scientific evidence for them, and in clinical trials reiki is no more effective than fake therapy[2] (the same is true for homeopathy),[3] so I find it hard to believe that the treatment has any direct physical effects. Yet many people – like Sue – clearly feel that they benefit from reiki and other alternative therapies, spending millions of dollars on them, despite the damning trial results. Something is helping these patients, and I'm curious to know whether that something is the mind. So I'm here to see what reiki can do for me.

My session doesn't start too well. Our energy field radiates beyond our bodies, says McRoberts, and we can see it if we look hard enough. All I see, however, is McRoberts and the wall. She adjusts the blind and I stare until my eyes blur. 'It's hard to tell,' I say hesitantly, not wishing to offend so early on. My therapist is undaunted. Children are better at it, she shrugs, and we get to work.

McRoberts is warm and smiley, with a taut, tanned face and flowing scarves. Today, she informs me, she'll combine reiki with psychic healing. She calls on her spirit guides and helpers to join us, then calls on mine too. It doesn't matter if you believe, she says softly. They're coming anyway. She lays one hand on my tummy and raises the other, her fingers making flicking, darting movements in the air above my body.

My energy is closed, she says, hard and smooth like the under-side of a fibreglass boat. To soften it, she asks me to breathe deeply and relax. Her voice is soothing, I'm snug under the blankets, and somewhere there's the gentle sound of trickling water. I start to feel numb, limbs tingling, as if I'm floating. Then McRoberts gets a vision of me as a child, 'all knees and elbows', shouting something about not being listened to. She asks if that makes sense, but although I undoubtedly experienced moments of frus-tration growing up I was small and tubby as a child, never gangly, and I'm pretty sure I always made myself heard.

McRoberts asks if anyone close to me has 'crossed', and I say my grandpa. I can guess what will happen and, sure enough, she tells me he's in the room. Did he used to say something about a cork, she asks? 'Put a cork in a bottle . . . No, that's not quite it.' I wonder if I'm supposed to jump to the obvious phrase, 'Put a cork in it', but I don't remember my grandpa ever saying this, so I stay quiet.

What about my father, she asks, and I say no, he's alive. She says she can see him too, she's getting a vision of a man's legs: one crossed over the other, trousers creased, foot tapping. She repeats the image, of someone severe, judgemental and unforgiving. Perhaps this is how she imagines British fathers, but I don't recognise it – and by now I'm feeling bad about continually disappointing her.

McRoberts moves up to my head, fingers pressing on my forehead and the base of my skull, massaging behind my ear. I don't have any serious physical conditions, but McRoberts diagnoses fear. It's bubbling in your chest, she tells me. You're afraid that if you let go, everything will fall apart. This makes some sense; as a working mum I feel I have a lot to juggle. I'd call it stress, but McRoberts says it's fear that stems from not receiving unconditional love as a child.

She asks if I'm married. No, I say, but I live with my partner. I don't mention the children because she doesn't ask, and if McRoberts really can see my aura it does not appear to reveal to her this crucial part of my identity (she tells me later that she would not expect to see someone's children 'unless that is where the healing needs to occur'). There are serious problems in my relationship too, she warns, and a decision to make. Apparently I'm repeating the mistakes of my past, and deserve to be with someone who loves me no matter what. Rather than a settled mother-of-two, I wonder if she has me down as waiting in vain for a no-good boyfriend to propose.

Time for the healing. McRoberts vigorously moves her hands up and down over my body, telling me she's opening a channel of energy down my spine to release the fear and pain I've been storing. Then she warns me to expect 'a profound shift in consciousness'. It doesn't matter if you believe it or not, she says. Your body will do the work.

***

I started this book in the park on a summer's day, considering whether, by harnessing the power of the mind, alternative treatments can offer something that conventional medicine has missed.

Twelve chapters later, I've learned how our brains control many aspects of our physiology, including the tools that the body has available – from hormones and natural painkillers to the weapons of the immune system – to ease symptoms and fight disease. Instead of responding purely to physical circumstances, I've seen how the brain uses our perception of our environment, including memories of the past and predictions about the future, to decide how best to allocate its resources. These processes can have an effect within seconds, or they can influence our physiology for years to come.

We can rarely deploy these tools at will; we can't simply 'wish' ourselves better. But as described in these pages, there are ways in which we can use our conscious minds to influence them, from believing we have taken a pill, or focusing on the present moment, to seeking the support of someone we love.

At the heart of almost all the pathways I've learned about is one guiding principle: if we feel safe, cared for and in control – in a critical moment during injury or disease, or generally throughout our lives – we do better. We feel less pain, less fatigue, less sickness. Our immune system works with us instead of against us. Our bodies ease off on emergency defences and can focus on repair and growth.

What does this mean for alternative medicine? My reiki session hasn't persuaded me of the power of healing energy fields (not to mention friendly ghosts). But after learning about all the different ways in which the mind can affect the body, I can see that even if their treatments don't work in the way that they claim, therapists like McRoberts may still deliver a powerful blend of the healing elements described in this book.

As well as providing a caring one-on-one consultation with lashings of placebo effect, for example, McRoberts induced a relaxed state that to me felt very similar to hypnosis, including positive suggestions and dramatic visual imagery. My promised shift in consciousness did not occur, but for someone more hypnotisable or with more faith in her technique, I can believe that her approach might well reduce stress, and ease pain or fatigue more effectively than conventional drugs.

In trials too, alternative treatments can be very effective despite performing no better than placebo. In 2001, for example, Exeter University's Edzard Ernst carried out a rigorous trial of faith healing – a technique similar to reiki – for treating chronic pain.[4] He compared real therapists with actors (who had no training as healers, and silently counted backwards during the session to avoid inadvertently directing any healing thoughts towards the patient). There was no difference between the real and fake therapy but patients in both groups improved dramatically, with some, Ernst said afterwards, who 'practically abandoned their wheelchairs during the study'.[5]

So should we embrace alternative medicine? And should we care how it works, as long as it does?

One problem, of course, is that patients who adopt alternative medicines do not always have a positive outcome. While researching this book, for example, I met 37-year-old Tunde Balogh. Originally from Hungary, she lives in Ireland with her husband and young son. She's beautiful, with delicate, expressive features and sleek, brown hair; inside, though, her body is full of pain and riddled with disease. A year earlier, she had been diagnosed with cancer in her right breast. She refused conventional medical treatment. 'I was so against doctors, hospitals, nurses,' she told me. 'They offered me radiation. They would give me chemo. Or cut off my breasts. I didn't want to do that.'

Instead, she tried reiki, then reflexology. 'I knew inside – if I caused this I can fix this.' Then she found German New Medicine, which teaches that cancer is caused by emotional conflict; if we resolve the conflict, the cancer will be cured. Founder Ryke Hamer claims that women develop breast cancer when they are conflicted regarding their loved ones, or their role as a mother.[6] Tunde says this resonated with her, because insecurities about her body had been causing her to distance herself from her husband. 'Why did you do that, now you have cancer!' she says. 'It took me around six months to forgive myself.'

But her cancer wasn't cured. In January 2014, she started suffering from searing joint pains; the disease had spread to her bones. 'Cancer in the bones is when you don't feel valuable,' she says. She stood in front of the bedroom mirror each day, repeating: 'I'm valuable. I love myself.'

By June, Tunde struggled to walk and was in severe pain. She was as convinced as ever that the answer was inside her, though, and still searching for a cure. I met her in Lourdes, where she was on the same pilgrimage as Rose and her disabled children. She had washed her breast in the holy water and visited the grotto in a wheelchair, but had yet to go to the baths. Why come to Lourdes, I asked, if you believe healing is inside? Maybe as a confession, she said, for having caused her own cancer. 'Maybe to wash out my sin.'

It's important to remember that just because the mind plays a role in health, this does not mean it can cure everything, or that any therapy that might harness the mind is suddenly justified. Breast cancer generally has a good prognosis if treated early but is not curable once – as in Tunde – it has spread to the bones. When people reject conventional treatments in favour of untested ones, they may die as a result.

Tunde's case is extreme, perhaps, but there are plenty of reported cases of people who have died after rejecting conventional treatment in favour of alternative therapy.[7] And less dramatic examples can still risk lives. In 2002, British researchers approached 168 homeopaths and found that nearly half of them advised patients against the measles, mumps and rubella (MMR) vaccine for their children.[8] Similarly, a 2006 investigation by the BBC programme

Newsnight reported that almost all homeopaths they approached advised travellers against conventional drugs to protect against malaria, and recommended ineffective homeopathic remedies instead.[9] A homeopath in one high-street pharmacy told *Newsnight*'s researcher: 'They make it so your energy doesn't have a malaria-shaped hole in it so the malarial mosquitos won't come along and fill that in.' I find it hard not to feel angry reading such nonsens-ical – and potentially fatal – advice.

Physical complications resulting from alternative medicines are rare but do exist. Acupuncture needles have caused nasty infec-tions, for example,[10] and unlicensed herbal medications can have serious side effects. Another concern is the psychological harm that therapists might do to vulnerable patients. Tunde's physical deterioration is heart-breaking enough, but on top of that is the guilt she feels from believing that she caused her own cancer. Poorly trained hypnotherapists can unwittingly plant false memo-ries, for example of abuse. During my reiki session, when McRoberts told me I'm in pain because I'm not getting the love I need, her words did not hit home. But if I were seriously ill and desperate enough for a cure, might her treatment instead have caused me to turn against those close to me, and to blame them for my condition just when I most needed their support?[11]

There are increasing efforts to integrate conventional and alter-native treatments, ranging from individual GPs like Patricia Saintey – who offers therapies including homeopathy at her private clinic – to big hospitals. The NHS-funded Centre for Integrative Care in Glasgow offers holistic interventions such as homeopathy and mistletoe therapy, for example, while at Stanford Center for Integrative Medicine in the US, cancer patients can have traditional acupuncture alongside their chemotherapy. This helps to ensure that treatments given are regulated, and that patients also get any conventional care they need.

When I visited the Stanford Center, therapist Deming Huang explained how his needles 'adjust the energy function' of the body, and talked me through the 12 major energy channels – or meridians – that acupuncture targets. Western scientists haven't been able to find any evidence for these channels,[12] and the data regarding the benefits of the therapy are debated. Sham acupuncture – in which the needles don't penetrate the skin, or are used in the wrong place – generally has very similar effects in trials to the real thing (but both are significantly better than no treatment at all), suggesting that for most complaints, any benefit from acupuncture results from a potent placebo effect. Rigorous analyses suggest that it is slightly better than placebo, however, when used to treat nausea and some types of chronic pain.[13]

Huang performs acupuncture on cancer patients to help reduce the side effects of their treatment. 'Most of our patients have only mild symptoms,' he says. 'They are able to go through the whole treatment more smoothly.' That improves the survival rate, he claims, because more patients are able to finish their full treatment course. And it reduces costs, because when patients suffer side effects they visit him instead of their oncologist. 'They can see us four or five times for the cost of one oncologist appointment.'[14]

It's a controversial approach. Steven Salzberg, a computational biologist at the University of Maryland, College Park, and a prominent critic of alternative therapies, has described integrative medicine as 'cleverly marketed, dangerous quackery' and argues that treatments such as acupuncture should not be offered in publicly-funded medical centres.[15] Jeremy Howick, a philosopher of science and epidemiologist at the Centre for Evidence-Based Medicine in Oxford, disagrees. He argues that we shouldn't worry so much about understanding whether alternative therapies work through a physical or a psychological mode of action (or both), and that instead we should focus on how they compare in trials to established treatments. 'I think it's more important to know *that* something works, than *how* it works,' he says. 'If I had cancer, I wouldn't care what explanation the therapist gave. I would want him to cure my pain. Wouldn't you?'[16]

I probably would. But I'm still a little uneasy when, in offering alternative treatments, conventional doctors seemingly endorse explanatory frameworks with no scientific basis. It feels to me like an admission of defeat; a confession that these exotic explanations hold some power that science cannot unlock. Is it then surprising if people start to believe in the energy fields and auras that therapists tell them are responsible for the improvements they experience (not to mention healing spirits, German New Medicine, or whatever else comes along), or that they lose confidence in drugs and vaccines that really can save lives?

**

As individuals, rather than putting our faith in mystical rituals and practices, the science described in this book shows that in many situations, we have the capacity to influence our own health, by harnessing the power of the (conscious and unconscious) mind. If you feel that alternative remedies work for you, I don't see any need to abandon them, especially when conventional medicine does not yet provide all of the same elements. But be critical of the advice that you may be offered by alternative therapists. And give your brain and body some credit. It's not necessarily the potions or needles or hand waving that make you feel better. Consider the possibility that these are just a clever way of hitting your buttons, enabling you to influence your own physiology in a way that eases your symptoms and protects you from disease.

And when it comes to medicine, rather than importing alternative therapies wholesale, many of the scientists and doctors we've heard about are trying a different approach. They want to understand instead the true active ingredients of these treatments (such as empathy, social support, hope) and how to incorporate those into better patient care.

More basic research is needed; we're only just starting to understand the complexities of the connections between brain and body.

One intriguing area for research, for example, is whether men and women respond differently to stress. Studies so far suggest that men are more sensitive to achievement challenges such as mental arithmetic tasks or public speaking, while women are most vulnerable to interpersonal problems like social rejection.[17] 'We are very different animals,' concludes UCSF stress researcher Elissa Epel.[18] She's keen to know if this can help explain why men and women suffer from different patterns of stress-related disease, with men more susceptible to cardiovascular disease and diabetes, and women at higher risk of anxiety disorders and depression.

And we need more clinical trials to work out what actually helps patients in the real world. Even for one of the best-studied techniques, mindfulness, researchers need to test, for example, whether it works better in some populations than others; how it stacks up against the best available drugs for various conditions; and whether it extends beyond psychological benefits to reduce the biological impact of stress on the body and reduce disease risk long-term.

Already, though, we've seen lots of examples of researchers using some of the principles described in this book to change how patients are cared for, with striking results. They include Vicki Jackson, talking to terminally ill patients about what a good life means to them; Ted Kaptchuk, dispensing honest placebos; Elvira Lang, changing the language radiologists use when they speak to patients; and Hunter Hoffman, designing virtual worlds that melt away pain. All are combining holistic care with a rigorous approach to evidence. All are managing to reduce reliance on drugs and other physical interventions, while improving outcomes for patients.

There are of course countless other examples I haven't had space to describe in detail. Jeff Sloan, a health sciences researcher at the Mayo Clinic in Rochester, Minnesota, wants to help doctors take account of how patients *feel*, instead of relying solely on physical tests. That's tough to do in a rushed appointment. 'In modern medicine doctors usually only have one-to-three minutes of any given clinical visit with a patient that are unaccounted for,' he says. 'The rest of the time is tied up with doing the physical, or looking at lab tests and discussing the results.'[19]

So every oncology patient at the Mayo clinic is given three simple questions when they check in – they're asked to rate their pain, fatigue and quality of life on a scale from one to ten. Sloan says even this simple intervention is helping doctors to act on problems they might otherwise miss. Quality of life, for example, might sound like a vague psychological measure but it turns out to be crucially important for physical survival. 'We know that if you score five or less on that question, the risk of you dying from your cancer doubles,' says Sloan.[20]

A growing network of buildings in the UK called Maggie's centres offers a very different approach, yet still guided by the importance of patient experience. Intended as places where people with cancer can go for practical, emotional and social support, they aim above all to 'lift the soul'. The centres are designed by leading architects (including Frank Gehry and Zaha Hadid) to be welcoming, homely, intimate, beautiful – the opposite of many conventional hospitals. Visitors can chat to other patients; consult an oncology nurse or psychologist; get advice on nutrition or money; or just sit in the garden with a cup of tea.

I'm not aware of any randomised controlled trials comparing how patients who visit Maggie's centres fare compared to others. But as one advocate argues in the *British Medical Journal*, 'if any of these buildings contributes to a pleasantly thoughtful or reflective moment for any of its users, a moment with friends or relatives, or a moment of hope and calm that they otherwise wouldn't have had then they have already achieved something wonderful.'[21]

***

This is the point at which I'd love to conclude that thanks to studies and projects like these we are witnessing a revolution in medicine, in which we'll soon fully understand the role of the mind in health and will come to see the human aspects of care not as an add-on luxury, but as a central, guiding principle towards improving patient outcomes. Unfortunately, the odds are stacked against that happening.

One obstacle is the way in which research is funded: more than three quarters of clinical trials in the US are funded by drug companies,[22] who understandably have no interest in proving the benefit of any approach to care that might reduce the need for their products. Pills and medical devices are clearly a more attractive business proposition than hypnotherapy or biofeedback. The enthusiasm for physical interventions goes beyond market forces, however: almost all public money goes on conventional drug research too. The annual budget of the US National Institutes of Health (NIH) is around $30 billion, for example, of which less than 0.2% goes on testing mind–body therapies.[23]

The bigger problem, I'd argue, is a wider, deep-seated prejudice against the idea that the mind might have the power to heal, or to keep us well. The materialist world view described in this book's introduction – which prioritises physical test results and interventions, and sees subjective experience as a distraction – still reigns supreme in science. (Sloan recalls that when he carried out a study showing that some terminally ill cancer patients in palliative care rate their quality of life just as high as healthy people, the first response of reviewers was that 'the patient must be wrong'.) Ignoring subjective experience is great when you're trying to eliminate bias from your scientific experiments but is not always helpful in caring for patients, when psychological and physical wellbeing are inextricably entwined.

Western medicine is (rightly) underpinned by science and trial evidence, and to many policy-makers and funders, physical interventions just 'feel' more scientific than mind–body approaches do. Bioelectronics researcher Kevin Tracey is now enjoying millions of dollars of private and public funding to pursue his idea of stimulating the nervous system with electricity, even though as I write this, his largest published human study is in eight people. Gastroenterologist Peter Whorwell, by contrast, can't persuade local funding agencies to pay for his IBS patients to receive gut-focused hypnotherapy despite decades of positive trials in hundreds of patients.

'I think there is a double standard,' says Howick, of the Centre for Evidence-Based Medicine. 'A common stick used to beat non-conventional trials is that they're lower quality,' he says. 'It's not true.' Mindfulness has been subjected to hundreds of well-designed trials, he says. A 2005 analysis of 110 homeopathy trials found that they were of higher quality than equivalent studies of conventional drugs.[24]

This ingrained resistance to mind–body interventions is something I've heard about over and over again while researching this book. Even when scientists have funding, they often have to fight the surrounding culture in hospitals and universities just to conduct a trial.

Elvira Lang told me how the local ethics committee at Harvard responded to her plans to study patients undergoing keyhole surgery. 'I remember a time when I had two trials pending for the committee,' she says. 'One trial was on reading a script to patients to relax during the procedure. The other was carotid artery stenting in the very early days, where the way that trial was designed you had a pretty good chance of killing some people. The carotid trial was approved in no time! But the hypnosis trial, it took forever.'[25]

Meanwhile perinatal nursing expert Ellen Hodnett met resistance when trying to test whether women suffer fewer complications giving birth in an 'ambient' environment – with dim lighting, projected scenes from nature and a mattress on the floor – compared to a conventional hospital room, dominated by technical equipment and a bed. Most hospitals she approached flatly refused to make the requested changes, she says, even though the medical devices would still have been close by. 'Anybody who takes this on has an awful lot to overcome in terms of provider beliefs and attitudes to even allow the trial to go forward.'[26]

In a medical system based on evidence from trial results, the medicine we end up with depends on the trials that are carried out. So perhaps it's not surprising that in western medicine, there is little attempt to nurture and harness patients' psychological resources. Despite their best intentions, medical professionals are working within a system that prioritises access to medical technology and allows increasingly little space for the human aspects of care.

In the US, 'physicians have become part of an assembly line of care', says Bill Eley, associate dean of Emory University School of Medicine in Atlanta, Georgia. 'We are increasingly pushed to see more patients in less time.'[27] It's a trend he fears is contributing to a loss of empathy among medical professionals (and in turn to scary rates of depression and burnout).[28] Appointment times are squeezed to cut costs, even though the country spends nearly $3 *trillion* a year on healthcare; that's more than 17% of GDP, higher than anywhere else in the world.[29] Meanwhile prescription drug use is dizzyingly high. Almost half of all Americans are on medication,[30] most commonly for cardiovascular disease and high cholesterol (both of which are influenced by stress) with nearly 60% of adults aged over 65 taking five or more different drugs at any one time (18% are on at least ten).[31]

Of course, physical interventions – from drugs to heart surgery – are crucially important. When my baby son suffered a lung infection, the antibiotics he received quite possibly saved his life, and I couldn't have cared less about his consultant's bedside manner. The ability to cure and prevent childhood infections, in particular, is a gift those of us in developed countries are now fortunate enough to take for granted.

But the main threats facing us now are not acute infections, easily cured with a pill, but chronic, stress-related conditions for which drugs are not nearly as effective. We've seen that in many cases, painkillers and antidepressants may not work much better than placebo. The top ten highest grossing drugs in the US help only between 1 in 25 and 1 in 4 of the people who take them; statins may benefit as few as 1 in 50.[32]

Meanwhile medical interventions are causing harm that dwarfs any damage done by alternative treatments. In 2015, an analysis of psychiatric drug trials published in the *British Medical Journal* concluded that these drugs are responsible for more than half a million deaths in the western world each year, in return for minimal benefits.[33] Medical errors in hospital are estimated to cause more than 400,000 deaths per year in the US alone – making it the third leading cause of death after heart disease and cancer – with another 4–6 million cases of serious harm.[34] According to the US Food and Drug Administration, there are another 2 million serious cases of adverse drug reactions in the US each year including 100,000 deaths.[35]

These statistics don't include expected side effects and complications of medications and interventions (many of which, as we heard in Chapter 7, people might not need under a different model of care), or the huge problems caused by prescription drug abuse, for example, or the rise of antibiotic resistance. The United States is the richest country in the world, yet even with trillions of dollars to spend it cannot match the life expectancy of a middle-income country like Costa Rica.

I am not advocating relying solely on the mind to heal us; but denying its role in medicine surely isn't the answer either. My hope, then, is that this book might help to overcome some of the prejudice against mind–body approaches, and to raise awareness that taking account of the mind in health is actually a *more* scientific and evidence-based approach than relying ever more heavily on physical interventions and drugs.

Perhaps one day this realisation might help lead towards a

system of medicine that combines the best of both worlds: one that uses life-saving drugs and technologies when they are needed, but also supports us to reduce our risk of disease and to manage our own symptoms when we are ill; and when there is no cure, cares for us and allows us to die with dignity. I hope that such a system of medicine would respect patients as equal participants whose beliefs, experiences and preferences matter in their care; that it would no longer stigmatise those with unexplained symptoms; and that it would recognise that the vast majority of health problems we face aren't physical or psychological – they are both.

The problems with modern medicine run deep; clearly they won't all be solved by mind–body therapies. But trying to improve medical outcomes by treating patients as the complex human beings we are, rather than simply as physical bodies, seems to me to be not such a bad place to start.

***

The implications of embracing the role of the mind in health go beyond medicine, of course. For me, one of the most surprising – and shocking – revelations of the research described in this book was that the stresses of poverty and inequality are sentencing large sections of the population to life-long chronic disease before they're even out of nappies. It's hard to disagree with the researchers arguing for social policies that aim to reduce those inequalities and in particular to support disadvantaged women of childbearing age. Meanwhile at the other end of the lifespan, projects like Experience Corps point to the potential of reframing ageing as a resource rather than a burden.

But there's one more insight that comes from understanding the links between mind and body. I've saved it until last because it's not just about health, medicine or society but something more fundamental. It tells us something about what it means to be human.

Ultimately, the science is saying that rather than passively experiencing the world around us, as most of us assume happens, to a large extent we construct and control that experience. 'Our bodies are not only receptors of information,' says placebo researcher Ted Kaptchuk. 'We create the information.' It's something that psychologists and neuroscientists are already discovering in other fields, such as memory and vision. Memories aren't faithful recordings but dynamic productions that we adapt and rewrite each time we access them, while our perception of colours and shapes is highly dependent on previous experience and what we expect to see.

Now it's clear this principle holds true for health too: our thoughts, beliefs, stress levels and world view all influence how ill or well we feel. As fatigue researcher Tim Noakes told us in Chapter 4, 'you don't have to believe what your brain is saying'.

The really new idea, here, though, is that when it comes to health, our minds determine far more than our subjective experience of the physical world around us. Through changes in gene expression, for example, and in the way our brains are wired, the way in which we see the world helps to shape our bodies too. We play a role, then, in constructing not just our experience but our *physical reality*. And in turn, the health of our physical bodies influences the state of our minds. Inflammation induces fatigue and depression. Low blood-sugar levels make us short-tempered.[36] Calming our bodies – by slow breathing, for example – improves our mood.

Nearly 400 years after Descartes' separation of the mind and body, we still tend to think of ourselves as logical, rational beings, with highly developed minds that allow us to transcend our biological, animal nature. The evidence shows something very different: that our bodies and minds have evolved in exquisite harmony, so perfectly integrated that it is impossible to consider one without the other. Terms like 'mind–body' and 'holistic' are often derided as flaky and unscientific, but in fact it's the idea of a mind distinct from the body, an ephemeral entity that floats somewhere in the skull like a spirit or soul, that makes no scientific sense.

This integration means we're not always as objective and reasonable as we might like to think. With our minds as well as our bodies shaped by evolution, we're built to hold beliefs that aid our health and survival, not that are necessarily true. There are powerful evolutionary forces driving us to believe in God, or in the remedies of sympathetic healers, or to believe that our prospects are more positive than they are. The irony is that although those beliefs might be false, they do sometimes work: they make us better.

By understanding how our minds influence and reflect our physiology, perhaps we can finally resolve that paradox – and live in tune with our bodies in a way that is based on evidence, not delusion.

NOTES

1. Nahin, R.L. et al. *National Health Statistics Reports*, no. 18, July 2009. Available at: https://nccih.nih.gov/sites/nccam.nih.gov/files/nhsrn18.pdf

   This report gives figures for use of complementary and alternative medicine (CAM) in 2007. It does not give figures for prayer. The previous report for 2002 did ask about prayer specifically for health reasons – it found that overall, 62% of adults had used some form of CAM (36% if prayer was not included).

   Barnes, P.M. et al. *National Health Statistics Reports*, no. 343, May 2004. Available at: http://www.cdc.gov/nchs/data/ad/ad343.pdf

   A report giving figures for 2012 was released in 2015, but did not include any cost data. With narrower definition than previous surveys, it found that 34% of adults had used CAM in 2012.

   Clarke, T.C. et al. *National Health Statistics Reports*, no. 79, 10 February 2015. Available at: http://www.cdc.gov/nchs/data/nhsr/nhsr079.pdf

2. *National Ambulatory Medical Care Survey: 2010 Summary*

Tables. Available at: http://www.cdc.gov/nchs/data/ahcd/namcs_summary/2010_namcs_web_tables.pdf
This figure is for 2010.

3. Silberman, S. *The Journal of Mind–Body Regulation* 2011; 1: 44–52
At the time of writing, homeopathy is still available on the NHS in some parts of the UK, see: http://www.nhs.uk/Conditions/homeopathy/Pages/Introduction.aspx#available [accessed 30 April 2015]

4. Dunn, P.M. *Archives of Disease in Childhood – Fetal and Neonatal Edition* 2003; 88: F441–F443

Chapter 1

1. Horvath, K. et al. *Journal of the Association for Academic Minority Physicians* 1998; 9: 9–15
Other sources for the story of secretin include 'Secretin Trials: A drug that might help, or hurt, autistic children is widely prescribed but is just now being tested' by Steve Bunk (*The Scientist*, 21 June 1999) and an open letter from Victoria Beck available at: https://groups.google.com/forum/#!topic/alt.support.autism/lnDCRgEwbJ4

2. A transcript of the *Dateline* programme on secretin is available at: http://psydoc-fr.broca.inserm.fr/fora/aut_for1.html

3. Telephone interview with Adrian Sandler, 7 February 2014.

4. Sandler, A.D. et al. *New England Journal of Medicine* 1999; 341: 1801–1806

5. The children in the secretin group went from 59 to 50; there was no statistically significant difference between the two groups.

6. Telephone interview with Bonnie Anderson, 20 May 2014. Now in her eighties, Bonnie can't remember the exact date, but she thinks it was in 2005.

7. Interview with Jerry Jarvik, University of Washington, Seattle, 7 May 2014.

8. Telephone interview with David Kallmes, 16 May 2014.

9. Kallmes, D.F. et al. *New England Journal of Medicine* 2009; 361: 569–79

10. Anon. *The Lancet* 1954; ii: 321

11. Sandler, A.D. et al. *New England Journal of Medicine* 1999; 341: 1801–1806

12. Huedo-Medina, T.B. et al. *British Medical Journal* 2012; 345: e8343

13. Hardy, J. et al. *Journal of Clinical Oncology* 2012; 30: 3611–3617

14. Wartolowska, K. et al. *British Medical Journal* 2014; 348: g3253

15. Rosanna spoke to me in Italian; her words were translated into English by Elisa Frisaldi.

16. de la Fuente-Fernandez, R. et al. *Science* 2001; 293: 1164–1166

17. 'The Power of the Placebo', *Horizon* BBC2, February 2014

18. Benedetti, F. et al. *Nature Neuroscience* 2004; 7: 587–588

19. See: http://www.redbullstratos.com/the-team/felix-baumgartner/

20. Interviews with Fabrizio Benedetti, Breuil-Cervina, 21 March 2014, and Plateau Rosa, 22 March 2014.

21. Levine, J.D., Gordon, N.C. & Fields, H.L. *The Lancet* 1978; 312: 654–657

22. Kirsch, I. *Epidemiologia e psichiatria sociale* 2009; 18: 318–322
    Kirsch, I. *The Emperor's New Drugs: Exploding the Antidepressant Myth* (Basic Books, 2011)

23. Benedetti, F., Carlino, E. & Pollo, A. *Clinical Pharmacology & Therapeutics* 2011; 90: 651–661

24. Wechsler, M.E. et al. *New England Journal of Medicine* 2011; 365:119–126

25. Chvetzoff, G. & Tannock, I.F. *Journal of the National Cancer Institute* 2003; 95: 19–29

26. Freed, C.R. et al. *New England Journal of Medicine* 2001; 344: 710–719

27. McRae, E. et al. *Archives of General Psychiatry* 2004; 6: 412–420

Chapter 2

1. Interview with Ted Kaptchuk, Cambridge, Massachusetts, 28 May 2014.
2. Kaptchuk, T.J., et al. *British Medical Journal* 2006; 332: 391
3. Moerman, D.J. *Medical Anthropology Quarterly* 2000; 14: 51–72
   According to Moerman, one of the major arguments for meaning as the source of placebo effects comes from the evidence for such cultural differences. Moerman has carried out extensive research on this topic, with many of the findings summarised in Chapter 6 of his 2002 book, *Meaning, Medicine and the Placebo Effect*.
4. Amanzio, M., Pollo, A., Maggi, G. & Benedetti, F. *Pain* 2001; 90: 205–215
5. Telephone interview with Dan Moerman, 20 April 2011, confirmed via email May 2015.
6. Walsh, B.T., Seidman, S.N., Sysko, R. & Gould, M. *Journal of the American Medical Association* 2002; 287: 1840–7
7. Kaptchuk, T.J. et al. *PLoS ONE* 2010; 5: e15591
8. Kelley, J.M., et al. *Psychotherapy & Psychosomatics* 2012; 81: 312–314
9. Kam-Hansen, S. et al. *Science Translational Medicine* 2014; 6: 218ra5
10. See: http://www.aplacebo.com/
11. Moerman, D. *Pain Practice* 2006; 6: 233–236
12. Email interviews with Edzard Ernst, 4 February 2014 and 13 April 2015.
13. See: http://edition.cnn.com/2012/05/29/world/asia/afghanistan-girls-poisoned/
14. *World Health Organization Weekly Epidemiological Monitor* vol 5, issue 22: Sunday 27 May 2012
15. Lorber, W., Mazzoni, G. & Kirsch, I. *Annals of Behavioral Medicine* 2007; 33: 112–116

Witthöft, M. & Rubin, G.J. *Journal of Psychosomatic Research* 2013; 74: 206–212

16. Reeves, R.R., Ladner, M.E., Hart, R.H. & Burke, R.S. *General Hospital Psychiatry* 2007; 29: 275–277

17. Silvestri, A. et al. *European Heart Journal* 2003; 24: 1928–1932

18. Humphrey postulates the existence of a 'health governor' in the brain, which acts like a hospital administrator, forecasting the body's future needs and allocating costly resources (from immune responses to self-generated symptoms, such as pain or fever) appropriately.

    These ideas are discussed in 'Great Expectations: The evolutionary psychology of faith healing and the placebo effect', an essay in Humphrey's 2002 book *The Mind Made Flesh* (pp. 255–285). A more recent review is Humphrey, N. & Skoyles, J. *Current Biology* 2012; 22: R1–R4.

19. Benedetti, F., Durando, J. & Vighetti, S. *Pain* 2014; 155: 921–928

20. This quote originally appeared in the article 'Heal Thyself' by Jo Marchant, *New Scientist*, 27 August 2011, pp. 30–34.

21. Walach advocates the use of alternative medicine, a view that in 2012 helped to win him a German sceptics' award for pseudoscience called the 'Goldene Brett'.

22. Walach, H. & Jonas, W.B. *Journal of Alternative and Complementary Medicine* 2004; 10: S–103-S–112

23. Telephone interview with Irving Kirsch 20 April 2011, confirmed via email May 2015.

24. Kaptchuk, T.J. et al. *British Medical Journal* 2008; 336: 999

25. Gracely, R.H. et al. *The Lancet* 1985; 1: 43

26. McMillan, F.D. *Journal of the American Veterinary Medical Association* 1999; 215: 992–999

27. Jensen, K.B. et al. *Proceedings of the National Academy of Sciences* 2012; 109: 15959–15964

Chapter 3

1. Someone with a transplanted kidney is two to three times more likely to develop cancer compared to a person of the same age and sex in the general population, mainly because the drugs that prevent their body from rejecting the organ also suppress immune responses that would normally protect them from cancer.
    Wong, G. et al. *Kidney International* 2014; 85: 1262–1264

2. Interview with Fabrizio Benedetti, Breuil-Cervina, 21 March 2014, and email interview 13 February 2014.

3. Telephone interview with Adrian Sandler, 7 February 2014.

4. Sandler, A.D. et al. *Journal of Developmental & Behavioral Pediatrics* 2010; 31: 369–375

5. Ader, R. & Cohen, N. *Psychosomatic Medicine* 1975; 37: 333–340

6. Interview with Manfred Schedlowski, University of Essen, 27 March 2014.

7. Vitello, P. *New York Times* 29 December 2011, p. B8

8. *Healing and the Mind with Bill Moyers* 1993, Ambrose Video Publishing, Vol 2: The Mind Body Connection

9. Williams, J.M. et al. *Brain Research Bulletin* 1981; 6: 83–94

10. *The Rochester Review*, 1997; vol 59, no 3. Available at: http://www.rochester.edu/pr/Review/V59N3/feature2.html

11. *Healing and the Mind with Bill Moyers* 1993, Ambrose Video Publishing, Vol 2: The Mind Body Connection

12. Ader, R. & Cohen, N. *Science* 1982; 215: 1534–1536

13. *Healing and the Mind with Bill Moyers* 1993, Ambrose Video Publishing, Vol 2: The Mind Body Connection.

14. Olness, K. & Ader, R. *Developmental and Behavioral Pediatrics* 1992; 13: 124–125

15. Giang, G.W. et al. *The Journal of Psychiatry & Clinical Neurosciences* 1996; 8: 194–201

16. Telephone interview with Karen Olness, 27 February 2014.

17. Exton, M.S. et al. *Transplantation Proceedings* 1998; 30: 2033

18. Exton, M.S. et al. *American Journal of Physiology – Regulatory, Integrative and Comparative Physiology* 1999; 276: 710–717

19. Vits, S. et al. *Brain, Behavior & Immunity* 2013; 29: S17

20. Goebel, M.U. et al. *Psychotherapy & Psychosomatics* 2008; 77: 227–234

21. This statistic comes from Witzke. For more detailed statistics, see: http://srtr.transplant.hrsa.gov/annual_reports/2012/

22. Interview with Oliver Witzke, University of Essen, 27 March 2014.

23. Ghanta, V.K. et al. *Annals of the New York Academy of Sciences* 1987; 496: 637–646
Ghanta, V.K. et al. *Annals of the New York Academy of Sciences* 1988; 521: 29–42
Ghanta, V.K. et al. *Cancer Research* 1990; 50: 4295–4299
Ghanta, V.K. et al. *International Journal of Neuroscience* 1993; 71: 251–265

24. Ader, R. et al. *Psychosomatic Medicine* 2010; 72: 192–197

25. Doering, B.K. & Rief, W. *Trends in Pharmacological Sciences* 2012; 33: 165–172

Chapter 4

1. West, J.B. *High Life: A History of High-Altitude Physiology and Medicine* (1998), Oxford University Press, p. 281

2. West, J.B. *High Life: A History of High-Altitude Physiology and Medicine* (1998), Oxford University Press, p. 282

3. Grocott, M.P.W. et al. *New England Journal of Medicine* 2009; 360: 140–149

4. The oxygen content of the air we breathe in falls as we climb, of course, but up to 7,100 metres – in these experienced, acclimatised climbers at least – the body was able to compensate for

this by increasing the amount of haemoglobin (the molecule that transports oxygen) in the blood.

5. Email interview with Dan Martin, 11 May 2015.

6. Noakes, T.D. *Journal of Applied Physiology* 2009; 106: 737–738

7. This is known in the field as 'the lactate paradox'. For a discussion of the evidence for this effect, see:
    West, J.B. *Journal of Applied Physiology* 2007; 102: 2398–2399
    Van Hall, G. *Journal of Applied Physiology* 2007; 102: 2399–2401
    West, J.B. *Journal of Applied Physiology* 2007; 102: 2401

8. BBC London 2012 coverage; clip available at:
    http://www.bbc.co.uk/sport/0/olympics/18912882

9. BBC London 2012 coverage; article available at:
    http://www.bbc.co.uk/sport/0/athletics/19230671

10. Nathan, M. et al. *South African Medical Journal* 1983; 64: 132–137
    Kew, T. et al. *South African Medical Journal* 1991; 80: 127–133
    Noakes, T. et al. *British Medical Journal* 1995; 310: 1345–1346

11. Noakes, T.D. *South African Medical Journal* 2012; 102: 430–432

12. Email interview with Tim Noakes, 22 April 2014.

13. St Clair Gibson, A. et al. *American Journal of Physiology – Regulatory, Integrative and Comparative Physiology* 2001; 281: R187–R196
    Kay, D. et al. *European Journal of Applied Physiology* 2001; 84: 115–121
    For more discussion of the evidence for Noakes' central governor, see the article 'Running on Empty' by Rick Lovett, *New Scientist*, 20 March 2004, pp. 42–45

14. Noakes, T.D. et al. *The Journal of Experimental Biology* 2001; 204: 3225–3234
    Noakes, T.D. *Applied Physiology, Nutrition and Metabolism* 2011; 36: 23–35

15. Email interview with Dan Martin, 18 May 2015.

16. Swart, J. et al. *British Journal of Sports Medicine* 2009; 43: 782–788

17. Okano, A.H. et al. *British Journal of Sports Medicine* 2013; doi:10.1136/bjsports-2012-091658

18. Beedie, C.J. & Foad, A. *Sports Medicine* 2009; 39; 313–329

19. Interview with Chris Beedie, London, 10 April 2014.

20. Pollo, A. et al. *European Journal of Neuroscience* 2008; 28: 379–388

21. Cairns, R. & Hotopf, M. *Occupational Medicine* 2005; 55: 20–31

22. This might be about to change, however. A 2015 study that analysed blood samples from nearly 650 people found that those who had been ill for less than three years had higher levels of chemicals that induce inflammation in the body compared to healthy controls, while those who had been sick for longer had lower-than-normal levels.
    Hornig, M. et al. *Science Advances* 2015; 1: e1400121

23. White, P.D. et al. *The British Journal of Psychiatry* 1998; 173: 475–481

24. For information about the trials, see:
    Edmonds, M. et al. *Cochrane Database of Systematic Reviews* 2004; 3: CD003200
    Bagnall, A.-M. et al. 'The Treatment and Management of Chronic Fatigue Syndrome (CFS)/Myalgic Encephalomyelitis (ME) in Adults and Children: Update of CRD Report 22'. Available at: http://www.york.ac.uk/media/crd/crdreport35.pdf
    Malouff, J.M. et al. *Clinical Psychology Review* 2008; 28: 736–45
    Price, J.R. et al. *Cochrane Database of Systematic Reviews* 2008; 3: CD001027

25. Telephone interview with Peter White, 2 May 2014.

26. White, P.D. et al. *The Lancet* 2011; 377: 823–836

27. *The Lancet* 2011; 377: 1808

28. Collings, A.D. & Newton, D. Response to White, P.D. *British Medical Journal* 2004; 329: 928. Available at: http://www.bmj.com/content/329/7472/928/rr/702549

29. Blackmore, S.J. Response to White, P.D. *British Medical Journal* 2004; 329: 928. Available at: http://www.bmj.com/content/329/7472/928/rr/759419

30. For more information on Samantha's art, please see: http://www.samantha-miller.co.uk/

Chapter 5

1. Interview with Peter Whorwell, Withington Community Hospital, Manchester, 14–15 May 2014.

2. Herr, H.W. *Urologic Oncology: Seminars and Original Investigations* 2005; 23: 346–351

3. Interview with David Spiegel, Curie Institute, Paris, 23 October 2013.

4. We vary in how hypnotisable we are. The classic scale of hypnotisability involves giving people a series of test suggestions that they pass or fail, for example that their arm will rise by itself, or that they'll see their best friend in the room. It's generally said that around 80% of the population score in the medium range, with 10% of people highly hypnotisable and 10% barely hypnotisable at all (for example, see hypnosis.tools/measurement-of-hypnosis.html). How people score on this test varies slightly in different studies and in different populations tested, however (for example, see Bongartz, W. *International Journal of Clinical and Experimental Hypnosis* 1985; 33: 131–139).

5. Kosslyn, S.M. et al. *The American Journal of Psychiatry* 2000; 157: 1279–1284

6. Dikel, W. & Olness, K. *Pediatrics* 1980; 66: 335–340

7. Telephone interview with Karen Olness, 27 February 2014.

8. Casiglia, E. et al. *American Journal of Clinical Hypnosis* 1997; 40: 368–375

9. Casiglia, E. et al. *International Journal of Psychophysiology* 2006; 62: 60–65

10. Casiglia, E. et al. *American Journal of Clinical Hypnosis* 2007; 49: 255–266

11. Email interview with Edoardo Casiglia, 4 March 2014.

12. For example:
   Kiecolt-Glaser, J.K. et al. *Journal of Consulting and Clinical Psychology* 2001; 69: 674–682
   Naito, A. et al. *Brain Research Bulletin* 2003; 62: 241–253

13. For example:
   Hewson-Bower, B. & Drummond, P.D. *Journal of Psychosomatic Research* 2000; 51: 369–377 (upper respiratory infections)
   Spanos, N.P. et al. *Psychosomatic Medicine* 1990; 52: 109–114 (warts)
   Results are mixed, however. Karen Olness carried out a trial of 61 children with warts, who received either hypnotherapy, standard treatment or no treatment. There was no significant difference between the three groups.
   Felt, B.T. et al. *American Journal of Clinical Hypnosis* 1998; 41: 130–137

14. Whorwell, P.J. et al. *The Lancet* 1984; 324: 1232–1234

15. Miller, V. & Whorwell, P.W. *International Journal of Clinical and Experimental Hypnosis* 2009; 57: 279–292

16. Calvert, E.L. et al. *Gastroenterology* 2002; 123: 1778–1785
   Miller, V. & Whorwell, P.W. *International Journal of Clinical and Experimental Hypnosis* 2009; 57: 279–292

17. Miller, V. & Whorwell, P.J. *International Journal of Clinical and Experimental Hypnosis* 2008; 56: 306–317
   Mawdsley, J.E. et al. *The American Journal of Gastroenterology* 2008; 103: 1460–1469
   Keefer, L. et al. *Alimentary Pharmacological Therapy* 2013; 38: 761–71

18. Gonsalkorale, W.M. et al. *Gut* 2003; 52: 1623–1629

19. Lea, R. et al. *Alimentary Pharmacology & Therapeutics* 2003; 17: 635–642

20. Chiarioni, G., Vantini, I., de Iorio, F. & Benini, L. *Alimentary Pharmacology & Therapeutics* 2006; 23: 1241–1249

21. Whorwell, P.J. et al. *The Lancet* 1992; 340: 69–72

22. For example, see:
    Lindfors, P. et al. *American Journal of Gastroenterology* 2012; 107: 276–285
    Moser, G. et al. *American Journal of Gastroenterology* 2013; 108: 602–609

23. Peters, S.L. et al. *Alimentary Pharmacology & Therapeutics* 2015; doi: 10.1111/apt.13202

24. See: http://www.nhs.uk/conditions/hypnotherapy/Pages/Intro duction.aspx [accessed 24 March 2015]

25. Interview with Jeremy Howick, Oxford, 20 April 2015.

26. According to the NIH's online search tool, projectreporter. nih.gov, the NIH is currently funding five research projects with 'hypnosis' or 'hypnotherapy' in the title (compared to 35 for 'mindfulness', for example).

27. Miller, V. et al. *Alimentary Pharmacology & Therapeutics* 2015; doi: 10.1111/apt.13145

Chapter 6

1. Sam Brown's story is told in 'Burning Man' by Jay Kirk, *GQ* magazine, February 2012. Available at: http://www.gq.com/ news-politics/newsmakers/201202/burning-man-sam-brown-jay-kirk-gq-february-2012

2. Hoffman, H.G. et al. *Annals of Behavioral Medicine* 2011; 41: 183–191

3. Pilkington, E. 'Painkiller Addiction: The plague that is sweeping the US', *The Guardian*, 28 November 2012. Available at:

http://www.theguardian.com/society/2012/nov/28/painkiller-addiction-plague-united-states

4. The American Society of Interventional Pain Physicians (ASIPP) Fact Sheet. Available at: https://www.asipp.org/documents/ASIPPFactSheet101111.pdf

5. 'Opioids Drive Continued Increase in Overdose Deaths', *CDC Press Release*, 20 February 2013. Available at: http://www.cdc.gov/media/releases/2013/p0220_drug_overdose_deaths.html See also 'Vital Signs: Overdoses of opioid prescription pain relievers – United States, 1999–2008', *Centers for Disease Control and Prevention Morbidity and Mortality Weekly Report* 2011; 60: 1487–1492. Available at: http://www.cdc.gov/mmwr/preview/mmwrhtml/mm6043a4.htm

6. Ahmed, A. 'Painkiller Addictions Worst Drug Epidemic in US History', *Al Jazeera America*, 30 August 2013. Available at: http://america.aljazeera.com/articles/2013/8/29/painkiller-kill-morepeoplethanmarijuanause.html

7. 'Aron Ralston Shares His Incredible Story of Survival'. Available at: https://www.youtube.com/watch?v=83nk6zmu5_o

8. Telephone interview with Hunter Hoffman, 7 May 2014.

9. Figure from interview with Sam Sharar, University of Washington Medical Center, 8–9 May 2014. See also Hoffman, H. et al. *Annals of Behavioral Medicine* 2011; 41: 183–191

10. Reviewed in Hoffman, H. et al. *Annals of Behavioral Medicine* 2011; 41: 183–191

11. Maani, C.V. et al. *Journal of Trauma and Acute Care Surgery* 2011; 71: S125–130

12. This quote appears in 'Burning Man' by Jay Kirk, *GQ* magazine, February 2012. Available at: http://www.gq.com/news-politics/newsmakers/201202/burning-man-sam-brown-jay-kirk-gq-february-2012

13. Esdaile's treatment of Gooroochuan Shah is described in *Hidden Depths: The Story of Hypnosis* (2002) by Robin Waterfield, pp. 196–197.

14. Interview with David Patterson, Seattle, Washington, 10 May 2014.

15. Patterson, D.R. et al. *The International Journal of Clinical & Experimental Hypnosis* 2004; 52: 27–38

16. Patterson, D.R. et al. *The International Journal of Clinical & Experimental Hypnosis* 2010; 58: 288–300

17. Barnsley, N. et al. *Current Biology* 2011; 21: R945–946

18. Moseley, G.L. *Neuroscience & Biobehavioral Reviews* 2012; 36: 34–46

19. Telephone interview with Candy McCabe, 19 December 2014.

20. McCabe, C. *Journal of Hand Therapy* 2011; 24: 170–179
    Preston, C. & Newport, R. *Rheumatology* 2011; 50: 2314–2315

21. Rothgangel, A.S. et al. *International Journal of Rehabilitation Research* 2011; 34: 1–13

22. Interview with David Spiegel, Curie Institute, Paris, 23 October 2013.

Chapter 7

1. 'Childhood, Infant and Perinatal Mortality in England and Wales', *Office for National Statistics Bulletin* 2012. Available at: http://www.ons.gov.uk/ons/dcp171778_350853.pdf

2. Waldenstrom, U. et al. *Journal of Psychosomatic Obstetrics & Gynecology* 1996; 17: 215–228

3. Olde, E. et al. *Clinical Psychology Review* 2006; 26: 1–16

4. In England in 2013/14, the rate of 'unassisted deliveries' (without induction, caesarean, instrumental delivery or episiotomy, but including pain relief such as epidurals) was 44.5%. http://www.birthchoiceuk.com/Professionals/index.html

5. Hodnett, E.D. et al. *Cochrane Database of Systematic Reviews* 2012; issue 10, article no. CD003766

6. Telephone interview with Ellen Hodnett, 10 March 2014.

7. Gibbons, L. et al. 'The Global Numbers and Costs of Additionally Needed and Unnecessary Caesarean Sections Performed Per Year: Overuse as a barrier to universal coverage', World Health Report 2010. Background Paper 30. Available at: http://www.who.int/healthsystems/topics/financing/healthreport/30C-sectioncosts.pdf

8. England statistics: http://www.birthchoiceuk.com/Professionals/index.html
US statistics: http://www.cdc.gov/nchs/fastats/delivery.htm

9. This is well established in animals. There's very little research on this in humans, but for example, see:
Lederman, R.P. *American Journal of Obstetrics & Gynecology* 1978; 132: 495–500
Lederman, R.P. *American Journal of Obstetrics & Gynecology* 1985; 153: 870–877

10. Hodnett, E.D. et al. *Journal of the American Medical Association* 2002; 288: 1373–1381

11. Brocklehurst, P. et al. *British Medical Journal* 2011; 343: d7400

12. Symon, A. et al. *British Medical Journal* 2009; 338: b2060
Babies in the independent midwife group were more likely to die, but the authors concluded this was because this group included significantly more 'high-risk' women with pre-existing medical conditions and complications. When the researchers excluded these cases from their analysis, the death rate in both groups was the same.

13. Olsen, O. & Clausen, J.A. *Cochrane Database of Systematic Reviews* 2012, issue 9. Art. No. CD000352.

14. 'New Advice Encourages More Home Births', *NHS Choices*, 13 May 2014. Available at: http://www.nhs.uk/news/2014/05 May/Pages/New-advice-encourages-more-home-births.aspx

15. My son was born on the morning of 18 October 2012. My midwives, Jacqui Tomkins and Elke Heckel, are from the

London Birth Practice (www.londonbirthpractice.co.uk). Tomkins has been chair of Independent Midwives UK (IMUK) since 2013, and in 2014 was named midwife of the year at the British Journal of Midwifery Awards for her work in securing insurance for self-employed midwives.

16. As I'd previously had a c-section, my second pregnancy was officially 'high-risk', because of the possibility that my scar from the previous surgery might rupture during delivery, with serious consequences for the baby and me. According to NHS guidelines, I should not have attempted to give birth at home. However, my partner and I researched the evidence on uterine rupture and concluded that in our case, the extra risk was very small. We decided – supported by the head of midwifery at my local hospital – that for us this risk was outweighed by the benefits of continuous care at home.

17. 'NICE Confirms Midwife-led Care During Labour is Safest for Straightforward Pregnancies', *NICE Press Release*, 3 December 2014. Available at: https://www.nice.org.uk/news/press-and-media/midwife-care-during-labour-safest-women-straightforward-pregnancies

18. Hodnett, E.D. et al. *Journal of the American Medical Association* 2002; 288: 1373–1381

19. 'The Cost of Having a Baby in the United States', *Truven Health Analytics Marketscan Study*, January 2013. Available at: http://transform.childbirthconnection.org/wp-content/uploads/2013/01/Cost-of-Having-a-Baby1.pdf

20. Skype video interview with Elvira Lang, 24 April 2014.

21. Lang, E.V. et al. *The Lancet* 2000; 355: 1486–1490
Lang, E.V. et al. *Pain* 2006; 126: 155–164
Lang, E.V. et al. *Journal of Vascular and Interventional Radiology* 2008; 19: 897–905

22. Lang, E.V. & Rosen, M.P. *Radiology* 2002; 222: 375–382

23. Lang's company is called Hypnalgesics (see www.hypnalgesics.com). Lang has also written two books about Comfort Talk

– *Patient Sedation Without Medication* (2011), which is aimed at medical professionals, and *Managing Your Medical Experience* (2014), written for patients.

24. Lang, E.V. *Journal of Radiology Nursing* 2012; 31: 114–119

25. Lang, E.V. et al. *Pain* 2005; 114: 303–309

26. Providing tools that patients can use to cope for themselves, rather than simply chatting or comforting them in other ways, seems crucial. In a trial of 201 patients having tumours destroyed using chemicals or an electric current, Lang included a control group who were given 'empathic care', which included avoiding negative language and swiftly responding to requests (Lang, E.V. et al. *Journal of Vascular and Interventional Radiology* 2008; 19: 897–905). These patients ended up far more anxious than those who received standard care. They needed more drugs, and suffered so many complications – things like falling oxygen levels, or a dangerous spike in blood pressure – that Lang had to stop the study early (patients in the Comfort Talk group, who were also read a relaxation script, did much better than standard care). Lang says the nurses in the empathic care group tried to comfort their patients – discussing their own experiences with illness, for example, or stroking a patient's forehead – and she thinks that this interfered with the patients' own coping efforts. This wasn't part of the intended intervention, but, 'Suddenly everyone in the room wanted to be extra nice,' she says, 'and sometimes patients just wanted to be left in peace.'

27. Lang, E.V. et al. *Academic Radiology* 2010; 17: 18–23

28. Temel, J.S. et al. *The New England Journal of Medicine* 2010; 363: 733–742

29. Telephone interview with Vicki Jackson, 16 December 2014.

30. Temel, J.S. et al. *The New England Journal of Medicine* 2010; 363: 733–742

Chapter 8

1. Telephone interview with Robert Kloner, 23 April 2013.
2. Kloner, R.A. et al. *Journal of the American College of Cardiology* 1997; 30: 1174–1180
3. Meisel, S.R. et al. *The Lancet* 1991; 338: 660–661
   Trichopoulos, D. et al. *The Lancet* 1983; 1: 441–444
   Suzuki, S. et al. *The Lancet* 1995; 345: 981
4. When Kloner looked for a spike in cardiac deaths in New York after the terrorist attacks of 11 September 2001, for example, he didn't find one. He suggests that this is because most of the people who were in direct danger and therefore might have suffered from this effect – those who were inside the two towers – perished anyway when the buildings collapsed.
5. More information on the Whitehall studies is available here: https://www.ucl.ac.uk/whitehallII
6. Bobak, M. & Marmot, M. *British Medical Journal* 1996; 312: 421–425
7. Dhabhar, F.S. et al. *Psychoneuroendocrinology* 2012; 37: 1345–1368
8. Glaser, R. & Kiecolt-Glaser, J.K. *Nature Reviews Immunology* 2005; 5: 243–251
   Cohen, S. et al. *Journal of the American Medical Association* 2007; 298: 1685–1687
9. Cohen, S. et al. *Proceedings of the National Academy of Sciences* 2012; 109: 5995–5999
10. Christian, L.M. et al. *Neuroimmunomodulation* 2006; 13: 337–346
    Godbout, J.P. & Glaser, R. *Journal of Neuroimmune Pharmacology* 2006; 1: 421–427
11. McDade, T.W. *Proceedings of the National Academy of Sciences* 2012; 109 supp 2: 17281–17288
12. Chung, H.Y. et al. *Ageing Research* 2009; 8: 18–30

13. Chida, Y. et al. *Nature Clinical Practice Oncology* 2008; 5: 466–475

    Heikkilä, K. et al. *British Medical Journal* 2013; 346: f165

14. Jenkins, F.J. et al. *Journal of Applied Biobehavioral Research* 2014; 19: 3–23

15. Sloan, E.K. et al. *Cancer Research* 2010; 70: 7042–7052 (breast cancer)

    Lamkin, D.M. et al. *Brain, Behavior & Immunity* 2012; 26: 635–641 (acute lymphoblastic leukaemia)

    Kim-Fuchs, C. et al. *Brain, Behavior & Immunity* 2014; 40: 40–47 (pancreatic cancer)

16. Lemeshow, S. et al. *Cancer Epidemiology, Biomarkers & Prevention* 2011; 20: 2273–2279

17. Blackburn's role in working out their function won her a share of the 2009 Nobel Prize in Physiology or Medicine.

18. Epel, E.S. et al. *Proceedings of the National Academy of Sciences* 2004; 101: 17312–17315

19. Sapolsky, R. *Proceedings of the National Academy of Sciences* 2004; 101: 17323–17324

20. For a review, see: Lin, J. et al. *Mutation Research* 2012; 730: 85–89

    There are also clues to how stress influences telomeres; in lab studies, the stress hormone cortisol reduces telomerase activity, while molecules involved in inflammation erode telomeres directly. This process seems to work in both directions – when the telomeres of immune cells get too short, they pump out chemicals that further boost inflammation, see: Rodier, F. & Campisi, J. *Journal of Cell Biology* 2011; 192: 547–556.

21. This quote first appeared in 'Can Meditation Really Slow Ageing?' by Jo Marchant published by Mosaic, 1 July 2014. Available at: http://mosaicscience.com/story/can-meditation-really-slow-ageing. (The section from paragraph 2 on p.163 to paragraph 3 on p.164 is adapted from this article.)

22. Cawthon, R.M. et al. *The Lancet* 2003; 361: 393–395

23. Armanios, M. & Blackburn, E.H. *Nature Reviews Genetics* 2012; 13: 693–704

24. Codd, V. et al. *Nature Genetics* 2013; 45: 422–427

25. Epel, E.S. et al. *Aging* 2009; 1: 81–88
    Zhao, J. et al. *Diabetes* 2014; 63: 354–362

26. 'Poor' is defined by the federal government's poverty thresholds – for example for a family of four (with two children) in 2014, this was defined as an annual income of less than $24,008. For more information on the economic challenges facing rural communities in black belt counties, see: Brody, G.H., Kogan, S.M. & Grange, C.M. (2012). 'Translating Longitudinal, Developmental Research with Rural African American Families into Prevention Programs for Rural African American Youth'. In V. Maholmes & R.B. King (eds), *Oxford Handbook of Poverty and Child Development*. London: Oxford University Press.

27. Telephone interview with Gene Brody, 8 January 2015, and interview, Emory University, Atlanta, 4 February 2014.

28. Brody, G.H., Kogan, S.M. & Grange, C.M. (2012). 'Translating Longitudinal, Developmental Research with Rural African American Families into Prevention Programs for Rural African American Youth'. In V. Maholmes & R.B. King (eds), *Oxford Handbook of Poverty and Child Development*. London: Oxford University Press

29. Miller, G.E. et al. *Psychological Bulletin* 2011; 137: 959–997

30. For example, see: http://www.ted.com/talks/richard_wilkinson?language=en

31. Telephone interview with Greg Miller, 4 December 2014. This research is summarised in Marmot, M. *The Status Syndrome: How Social Standing Affects Our Health and Longevity* (2005), Holt Paperbacks.

32. Miller, G.E. et al. *Proceedings of the National Academy of Sciences* 2009; 106: 14716–14721

33. Osler, M. et al. *International Journal of Epidemiology* 2006; 35: 1272–1277

34. Kittleson, M.M. et al. *Archives of Internal Medicine* 2006; 166: 2356–2361

35. Lin, J. et al. *Mutation Research* 2012; 730: 85–89

36. For example see:

    Szanton, S.L. et al. *International Journal of Behavioral Medicine* 2012; 19: 489–495

    Chae, D.H. et al. *American Journal of Preventive Medicine* 2014; 46: 103–111

    Brody, G.H. et al. *Child Development* 2014; 85: 989–1002

37. Blackburn, E.H. & Epel, E.S. *Nature* 2012; 490: 169–171

38. This quote (and the one in the following paragraph) first appeared in 'Can Meditation Really Slow Ageing?' by Jo Marchant published by Mosaic, 1 July 2014. Available at: http://mosaicscience.com/story/can-meditation-really-slow-ageing (Paragraphs 2–5 on p. 170 are adapted from this article.)

39. Telephone interview with Elissa Epel, 24 February 2014.

40. This concept (as well as the example with the skier) is described further in:

    Jamieson, J.P. et al. *Current Directions in Psychological Science* 2013; 22: 51–56.

41. Telephone interview with Wendy Mendes, 17 September 2014.

42. Jamieson, J.P. et al. *Current Directions in Psychological Science* 2013; 22: 51–56

43. Jamieson, J.P. et al. *Journal of Experimental Social Psychology* 2010; 46: 208–212

44. Chen, E. et al. *Child Development* 2004; 75: 1039–1052

45. Miller, G.E. et al. *Psychological Bulletin* 2011; 137: 959–997

46. McEwen, B.S. & Gianaros, P.J. *Annals of the New York Academy of Sciences* 2010; 1186: 190–222

    McEwen, B.S. & Morrison, J.H. *Neuron* 2013; 79: 16–29

47. Ganzel, B.L. et al. *NeuroImage* 2008; 40: 788–795

48. Miller, G.E. et al. *Psychological Bulletin* 2011; 137: 959–997

49. Sweitzer, M.M. et al. *Nicotine & Tobacco Research* 2008; 10: 1571–1575
50. Gianaros, P.J. et al. *Cerebral Cortex* 2011; 21: 896–910

Chapter 9

1. Paragraphs 1–2 and 18–19 of this chapter are adapted from 'Can Meditation Really Slow Ageing?' by Jo Marchant published by Mosaic, 1 July 2014. Available at: http://mosaicscience.com/story/can-meditation-really-slow-ageing

2. Telephone interview with Mark Williams, 9 February 2009, confirmed via email April 2015.

3. Pagnoni, G. et al. *PLoS One* 2008; 3: e3083

4. This quote is from Gareth Walker's video testimonial posted at: http://www.everyday-mindfulness.org/gareths-video-testimonial/ [accessed 2 April 2015]. All other quotes from Gareth Walker are from my interview, Barnsley, 23 January 2015.

5. Interview with Trudy Goodman, Santa Monica, 22 November 2013.

6. *National Health Statistics Reports*, no. 79, 10 February 2015. Available at: http://www.cdc.gov/nchs/data/nhsr/nhsr079.pdf

7. See Pickert, K. 'The Mindful Revolution', *TIME* magazine, 23 January 2014. Available at: http://time.com/1556/the-mindful-revolution/

8. For example, see:
Lauche, R. et al. *Journal of Psychosomatic Research* 2013; 75: 500–510
Lerner, R. et al. *Cancer and Clinical Oncology* 2013; 2: 62–72
Veehof, M.M. et al. *Pain* 2011; 152: 533–542
Piet, J. et al. *Journal of Consulting and Clinical Psychology* 2012; 80: 1007–1020

Hofmann, S.G. *Journal of Consulting and Clinical Psychology* 2010; 78: 169–183

Chiesa, A. & Serretti, A. *The Journal of Alternative and Complementary Medicine* 2011; 17: 83–93

Cramer, H. et al. *Current Oncology* 2012; 19: e343–351

9. For discussions of this see, for example:
    Blomfield, V. 'Buddhism and the Mindfulness Movement: Friends or foes?', blog post 6 April 2012. Available at: http://www.wiseattention.org/blog/2012/04/06/buddhism-the-mindfulness-movement-friends-or-foes/
    'Mindfulness: Panacea or fad?', BBC Radio 4, 11 January 2015. Presented by Emma Barnett. Produced by Phil Pegum. Available at: http://www.bbc.co.uk/programmes/b04xmqdd

10. Szalavitz, M. *Scientific American* July 2014: 30–31

11. Barker, K. *Social Science & Medicine* 2014; 106: 168–176

12. Interview with Gareth Walker, Barnsley, UK, 23 January 2015.

13. See: http://www.everyday-mindfulness.org/

14. Interview with Willem Kuyken, University of Exeter, 23 February. Since our meeting, Kuyken has moved to Oxford and is now director of the Oxford Mindfulness Centre.

15. Teasdale, J.D. et al. *Journal of Consulting and Clinical Psychology* 2000; 68: 615–623
    Ma, S.H. & Teasdale, J.D. *Journal of Consulting and Clinical Psychology* 2004; 72: 31–40
    These two randomised controlled trials compared MBCT with usual care, however they excluded patients currently taking antidepressants. Kuyken's subsequent trials of the therapy compared MBCT against drug treatment.

16. Kuyken, W. et al. *Journal of Consulting and Clinical Psychology* 2008; 76: 966–978

17. Kuyken, W. et al. *The Lancet* 2015; doi: 10.1016/S0140-6736(14)62222-4

18. Interview with Sara Lazar, Harvard University, Boston, 27 May 2014.

19. This quote previously appeared in 'Can Meditation Really Slow Ageing?' by Jo Marchant published by Mosaic, 1 July 2014. Available at: http://mosaicscience.com/story/can-meditation-really-slow-ageing

20. Lutz, A. *Proceedings of the National Academy of Sciences* 2004; 101: 16369–16373

21. Lazar, S.W. et al. *NeuroReport* 2005; 16: 1893–1897

22. Eriksson, P.S. et al. *Nature Medicine* 1998; 4: 1313–1317

23. Hölzel, B.K. et al. *SCAN* 2010; 5: 11–17
Hölzel, B.K. et al. *Psychiatry Research: Neuroimaging* 2011; 191: 36–43

24. Luders, E. *Annals of the New York Academy of Sciences* 2014; 1307: 82–88

25. Gard, T. et al. *Frontiers in Aging Neuroscience* 2014; 6: 76

26. Mohr, D.C. et al. *British Medical Journal* 2004; doi:10.1136/bmj.38041.724421.55

27. Buljevac, D. et al. *British Medical Journal* 2003; 327: 646

28. Mohr, D.C. et al. *Neurology* 2012; 79: 412–419

29. Results from the three-month meditation retreat studied by Blackburn and Epel are reported here: Jacobs, T.L. et al. *Psychoneuroendocrinology* 2011; 36: 664–681
Other examples of studies hinting that meditation might boost telomerase or lengthen telomeres include:
Ornish, D. et al. *The Lancet Oncology* 2013; 14: 1112–1120
Lavretsky, H. et al. *International Journal of Geriatric Psychiatry* 2013; 28: 57–65

30. This quote (and the quote from Elizabeth Blackburn in the following paragraph) previously appeared in 'Can Meditation Really Slow Ageing?' by Jo Marchant published by Mosaic, 1 July 2014. Available at: http://mosaicscience.com/story/can-meditation-really-slow-ageing

31. Interview with Elizabeth Blackburn, Paris, 23 October 2013.

32. Kabat-Zinn, J. et al. *Psychosomatic Medicine* 1998; 60: 625–632

33. Davidson, R.J. et al. *Psychosomatic Medicine* 2003; 65: 564–570
34. Barrett, B. et al. *Annals of Family Medicine* 2012; 10: 337–346
35. Simpson, R. et al. *BMC Neurology* 2014; 14: 15
36. Telephone interview with Robert Simpson, 7 January 2015.

Chapter 10

1. Rosero-Bixby, L. 'Costa Rican Nonagenarians: Are they the longest living male humans?' Paper presented at the IUSSP V International Population Conference, Tours, France, 2005
2. Rosero-Bixby, L. et al. *Vienna Yearb. Popul. Res.* 2013; 11: 109–136
3. Dan Buettner describes the visit in his 2010 book, *Blue Zones: Lessons for Living Longer From the People Who've Lived the Longest*, published by the National Geographic Society.
4. Rehkopf, D.H. et al. *Experimental Gerontology* 2013; 48: 1266–1273
5. Telephone interview with Michel Poulain, 2 September 2013.
6. House, J.S. et al. *American Journal of Epidemiology* 1982; 116: 123–140
7. House, J.S. et al. *Science* 1988; 241: 540–545
8. Holt-Lunstad, J. et al. *PLoS Medicine* 2010; 7: e1000316
9. Telephone interview with Charles Raison, 30 March 2011, confirmed via email May 2015. This quote originally appeared in the article 'Heal Thyself' by Jo Marchant, *New Scientist*, 27 August 2011, pp. 30–34. When we spoke, Raison was a professor at Emory University in Atlanta, Georgia. He is now based at the University of Wisconsin–Madison.
10. Vespa, J. et al. *America's Families & Living Arrangements: 2012* www.census.gov/prod/2013pubs/p20-570.pdf
11. McPherson, M. et al. *American Sociological Review* 2006; 71: 353–375

12. Eisenberger, N.I. et al. *Science* 2003; 302: 290–292
    Eisenberger, N.I. & Cole, S.W. *Nature Neuroscience* 2012; 15: 1–6

13. Cacioppo, J.T. et al. *Annals of the New York Academy of Sciences* 2011; 1231: 17–22
    Hawkley, L.C. & Cacioppo, J.T. *Annals of Behavioral Medicine* 2010; 40: 218–227

14. Telephone interview with John Cacioppo, 21 April 2011.

15. This quote originally appeared in the article 'Heal Thyself' by Jo Marchant, *New Scientist*, 27 August 2011, pp. 30–34

16. Luo, Y. et al. *Social Science & Medicine* 2012; 74: 907–914

17. Cole, S.W. et al. *Genome Biology* 2007; 8: R189

18. Interview with Steve Cole, University of California Los Angeles (UCLA), 21 November 2013.

19. Cole, S.W. et al. *Proceedings of the National Academy of Sciences* 2011; 108: 3080–3085

20. Cole, S.W. *PLoS Genetics* 2014; 10: e1004601

21. Antoni, M.H. et al. *Biological Psychiatry* 2012; 71: 366–372

22. Telephone interviews with Michael Antoni, 18 September 2013 and 6 March 2014.

23. This quote originally appeared in 'The Pursuit of Happiness' by Jo Marchant, *Nature* 2013; 503: 458–460

24. Spiegel, D. et al. *The Lancet* 1989; 334: 888–891

25. This was David Spiegel's count when I interviewed him at the Curie Institute, Paris, 23 October 2013. The negative trials include a large Canadian trial of 235 women with metastatic breast cancer, published in 2001 (Goodwin, P.J. et al. *New England Journal of Medicine* 2001; 345: 1719–1726), and Spiegel's own attempt to repeat his 1989 study, on 125 women with the condition, published in 2007 (Spiegel, D. et al. *Cancer* 2007; 110: 1130–7). Spiegel argues that there are problems with some of these studies, for example that the intervention being tested didn't cause any psychological changes in the first place, so wouldn't then be expected to have any physical effect.

The most prominent of the positive studies is a 2008 trial led by Barbara Andersen of Ohio State University, which included 227 women with non-metastatic breast cancer (Andersen, B.L. et al. *Cancer* 2008; 113: 3450–3458). They took a four-month course that aimed to provide them with social support and to help manage stress in their lives. Andersen followed the women for an average of 11 years. Their mood and immune responses improved, and their average survival time was increased by six months, from 2.2 years in the control group to 2.8 years in the therapy group. Sceptic James Coyne has criticised the statistical analysis used in this study, arguing that the data didn't actually show a positive result at all (Stefanek, M.E. et al. *Cancer* 2009; 115: 5612–5616).

26. Aizer, A.A. et al. *Journal of Clinical Oncology* 2013; 31: 3869–3876

For prostate, breast, colorectal, oesophageal and head/neck cancers, the authors concluded that the survival benefit conferred by being married was greater than that published for chemotherapy.

27. Interview with David Spiegel, Curie Institute, Paris, 23 October 2013.

28. Telephone interview with James Coyne, 19 September 2013.

29. Buchen, L. *Nature* 2010; 467: 146–148

30. McGowan, P.O. et al. *Nature Neuroscience* 2009; 12: 342–348

31. Lam, L.L. et al. *Proceedings of the National Academy of Sciences* 2012; 109: 17253–17260

Romans, S.E. et al. *Child Development* 2014; 86: 303–309

Naumova, O.Y. et al. *Development & Psychopathology* 2012; 24: 143–155

Fraga, M.F. et al. *Proceedings of the National Academy of Sciences* 2005; 102: 10604–10609

32. One of the first people to publish this idea was the biologist Bruce Lipton, in his 2005 book *The Biology of Belief: Unleashing the Power of Consciousness, Matter & Miracles*. It's

now a popular claim on new age and health websites, for example see:

http://www.abundance-and-happiness.com/epigenetics.html

http://healthscamsexposed.com/2014/06/epigenetics-proves-cancer-is-not-mysterious-or-inevitable/

http://healingthecause.blogspot.co.uk/2014/03/ancestral-healing-epigenetics.html

33. These ideas are discussed further in:

Cole, S.W. *Current Directions in Psychological Science* 2009; 18: 132–137

Cole, S.W. *PLoS Genetics* 2014; 10: e1004601

34. Brody, G.H., Kogan, S.M. & Grange, C.M. (2012). 'Translating Longitudinal, Developmental Research with Rural African American Families into Prevention Programs for Rural African American Youth'. In V. Maholmes & R.B. King (eds), *Oxford Handbook of Poverty and Child Development*. London: Oxford University Press.

Several other studies, for example by Northwestern University's Greg Miller, have also found that warm or nurturant parenting protects people against the biological effects of stress later in life.

Miller, G.E. & Chen, E. *Child Development Perspectives* 2013; 7: 67–73

35. Brody, G.H. et al. *Journal of Adolescent Health* 2008; 43: 474–481

36. Miller, G.E. et al. *Proceedings of the National Academy of Sciences* 2014; 111: 11287–11292

37. Telephone interview with Greg Miller, 4 December 2014.

38. Both loneliness and chronic stress are thought to increase the risk of dementia. For example, see:

Holwerda, T.J. et al. *Journal of Neurology, Neurosurgery and Psychiatry* 2014; 85:135–142

Greenberg, M.S. et al. *Alzheimer's & Dementia* 2014; 10: S155–S165

39. Telephone interview with Michelle Carlson, 24 February 2015.

40. Fried, L.P. et al. *Journal of Urban Health* 2004; 81: 64–78
    Carlson, M.C. et al. *Journal of Gerontology: Medical Sciences* 2009; 64: 1275–1282

41. Carlson, M.C. et al. *Alzheimers & Dementia*. Forthcoming.

42. Telephone interview with Lobsang Negi, 10 December 2014, and interview, Emory University in Atlanta, Georgia, 3 February 2015.

43. For more information on CBCT, see: http://tibet.emory.edu/cognitively-based-compassion-training/index.html

44. Pace, T.W.W. et al. *Psychoneuroendocrinology* 2009; 34: 87–98

45. Pace, T.W.W. et al. *Psychoneuroendocrinology* 2013; 38: 294–299

46. Mascaro, J.S. et al. *SCAN* 2013; 8: 48–55

47. Interview with Brendan Ozawa-de Silva, Atlanta, 4 & 5 February 2015.

Chapter 11

1. Novella, S. 'Energy Medicine: Noise-based pseudoscience', Science-based medicine blog, 12 December 2012. Available at: https://www.sciencebasedmedicine.org/energy-medicine-noise-based-pseudoscience/

2. The details of Janice's story (Janice is not her real name) given here are taken from the electronic version of Kevin Tracey's 2005 book *Fatal Sequence: The Killer Within*, published by Dana Press. Tracey notes in the introduction to this book that he did not take recordings or notes during Janice's hospitalisation, so he reconstructed the account from memory.

3. Levinson, A.T. et al. *Seminars in Respiratory and Critical Care Medicine* 2011; 32: 195–205

4. Tracey, K. *Fatal Sequence*, Chapter 5, location 1294

5. Tracey, K. *Fatal Sequence*, Introduction, location 70
6. Lehrer, P. *Biofeedback* 2013; 41: 88–97
7. Vaschillo, E. et al. *Applied Psychophysiology & Biofeedback* 2002; 27: 1–27
8. Lehrer, P. *Biofeedback* 2013; 41: 26–31
9. Thayer, J.F. & Lane, R.D. *Biological Psychology* 2007; 74: 224–242
10. Telephone interview with Paul Lehrer, 26 January 2015.
11. Del Pozo, J.M. et al. *American Heart Journal* 2004; 147: E11
    Lin, G. et al. *Journal of Alternative & Complementary Medicine* 2012; 18: 143–152
12. Gevirtz, R. *Biofeedback* 2013; 41: 110–120
13. Benson, H. *The Relaxation Response*, Avon Books, 1976, p. 83
14. For example, see:
    Benson, H. et al. *The Lancet* 1974; i: 289–291
    Benson, H. et al. *Journal of Chronic Diseases* 1974; 27: 163–169
15. Benson describes the results of his initial studies in his 1976 book, *The Relaxation Response* (pp. 87–95). For example, oxygen consumption abruptly dropped by 10–20% during meditation (compared to around 8% during sleep). Slow brain waves called alpha waves increased in intensity. Levels of lactic acid in the blood (a waste product of metabolism) dropped by around 40%. Heart rate slowed on average by about three beats per minute.
16. Park, G. & Thayer, J.F. *Frontiers in Psychology* 2014; 5: 278
    Porges, S.W. *Biological Psychology* 2007; 74: 116–143
17. Thayer, J.F. et al. *Neuroscience and Biobehavioral Reviews* 2012; 36: 747–756
18. Lehrer, P. *Psychosomatic Medicine* 1999; 61: 812–821
19. Gevirtz, R. *Biofeedback* 2013; 41: 110–120
20. Described in Tracey, K. *Fatal Sequence*, Chapter 7, location 1885
21. Described in Tracey, K. *Fatal Sequence*, Chapter 8, location 2307

22. Described in Tracey, K. *Fatal Sequence*, Chapter 9, location 2467

23. Watkins, L.R. et al. *Neuroscience Letters* 1995; 183: 27–31

24. Borovikova, L. et al. *Nature* 2000; 405: 458–462

25. Tracey, K.J. *Nature* 2002; 420: 853–859

26. Tracey tells this story in Tracey, K. 'Shock Medicine', *Scientific American* March 2015, pp. 28–35.

27. Kok, B.E. & Fredrickson, B.L. *Biological Psychology* 2010; 85: 432–436

28. Kok, B.E. et al. *Psychological Science* 2013; 24: 1123–1132

29. Telephone interview with Bethany Kok, 8 December 2014.

30. See: http://www.heartmath.com/science-behind-emwave/

31. These ideas are discussed in this interview with HeartMath's research director Rollin McCraty in 'Sufism: An inquiry' (vol 16, no 2, pp. 33–58). Available at: http://issuu.com/iasufism/docs/sufism.vol16.2

    See also:

    McCraty, R. et al. *The Journal of Alternative & Complementary Medicine* 2004; 10: 133–143

    McCraty, R. et al. *The Journal of Alternative & Complementary Medicine* 2004; 10: 325–336

    McCraty, R. & Childre, D. *Alternative Therapies in Health and Medicine* 2010; 16: 10–24

32. For example:

    Farkas, B. 'Is Heartmath's emWave Personal Stress Reliever Scientific?', James Randi Educational Foundation blog, 31 January 2011. Available at: http://archive.randi.org/site/index.php/swift-blog/1202--is-heartmaths-emwave-personal-stress-reliever-scientific-.html

    Novella, S. 'Energy Medicine: Noise-based pseudoscience', Science-based medicine blog, 12 December 2012. Available at: https://www.sciencebasedmedicine.org/energy-medicine-noise-based-pseudoscience/

33. Xin, W. et al. *American Journal of Clinical Nutrition* 2013; 97: 926–35

34. Video interview for Sky News. Available at: http://news.sky.com/story/1396464/nerve-hack-offers-arthritis-sufferers-hope

35. Koopman, F. A. et al. *Arthritis & Rheumatism* 2012; 64 Suppl 10: 581

36. Moore, T. '"Nerve hack" Offers Arthritis Sufferers Hope', Sky News, 23 December 2014. Available at: http://news.sky.com/story/1396464/nerve-hack-offers-arthritis-sufferers-hope

37. Tracey, K. 'Shock Medicine', *Scientific American* March 2015, pp. 28–35

38. Fritz, J.R. & Huston, J.M. *Bioelectronic Medicine* 2014; 1: 25–29

39. Miller, L. & Vegesna, A. *Bioelectronic Medicine* 2014; 1: 19–24

40. Behar, M. 'Can the Nervous System Be Hacked?', *New York Times* magazine, 23 May 2014. Available at: http://www.nytimes.com/2014/05/25/magazine/can-the-nervous-system-be-hacked.html

41. Martin, J.L.R. & Martín-Sánchez. E. *European Psychiatry* 2012; 27: 147–155

42. Behar, M. 'Can the Nervous System Be Hacked?', *New York Times* magazine, 23 May 2014. Available at: http://www.nytimes.com/2014/05/25/magazine/can-the-nervous-system-be-hacked.html

43. Weintraub, A. 'Brain-altering Devices May Supplant Drugs – and Pharma is OK With That', Forbes.com, 24 February 2015. Available at: http://www.forbes.com/sites/arlenewein-traub/2015/02/24/brain-altering-devices-may-supplant-drugs-and-pharma-is-ok-with-that/
Tracey, K. 'Shock Medicine', *Scientific American* March 2015, pp. 28–35

44. Guerrini, F. 'DARPA's ElectRx Project: Self-Healing Bodies through Targeted Stimulation of the Nerves', Forbes.com, 29

August 2014. Available at: http://www.forbes.com/sites/feder-icoguerrini/2014/08/29/darpas-electrx-project-self-healing-bodies-through-targeted-stimulation-of-the-nerves/

45. Tracey, K. *Fatal Sequence*, Chapter 10, location 2820

46. See, for example:
    Nolan, R.P. et al. *Journal of Internal Medicine* 2012; 272: 161–169
    Lehrer, P. et al. *Applied Psychophysiology and Biofeedback* 2010; 35: 303–315
    Kox, M. et al. *Psychosomatic Medicine* 2012; 74: 489–494
    Olex, S. et al. *International Journal of Cardiology* 2013; 18: 1805–1810

47. Behar, M. 'Can the Nervous System Be Hacked?', *New York Times* magazine, 23 May 2014. Available at: http://www.nytimes.com/2014/05/25/magazine/can-the-nervous-system-be-hacked.html

48. Tracey, K. *Fatal Sequence*, Chapter 10, location 2908

Chapter 12

1. Dawkins, R. *The God Delusion* (2006), Bantam Press
   Hawking, S. & Mlodinow, L. *The Grand Design* (2010), Bantam Press

2. 'Religion, Spirituality and Public Health: Research, applications and recommendations.' Testimony by Harold G. Koenig to Subcommittee on Research and Science Education of the US House of Representatives, 18 September 2008. Available at: https://science.house.gov/sites/republicans.science.house. gov/files/documents/hearings/091808_koenig.pdf

3. For example, a 2011 study of 36,000 adults in Norway found that the more often they attended church, the lower their blood pressure: Sorensen, T. et al. *The International Journal of Psychiatry in Medicine* 2011; 42: 13–28.

Another study of nearly 40,000 people in 22 countries found that those who went to church more reported better health: Nicholson, A. et al. *Social Science & Medicine* 2009; 69: 519–528. For a review, see Koenig, H.G. et al. *Handbook of Religion and Health* (2012), Oxford University Press.

4.   For example, see Sloan, R.P. et al. *The Lancet* 1999; 353: 664–667.

5.   'Religion, Spirituality and Public Health: Research, applications and recommendations.' Testimony by Harold G. Koenig to Subcommittee on Research and Science Education of the US House of Representatives, 18 September 2008. Available at: https://science.house.gov/sites/republicans.science.house. gov/files/documents/hearings/091808_koenig.pdf

6.   Telephone interview with Richard Sloan, 28 February 2015.

7.   Chida, Y. et al. *Psychotherapy & Psychosomatics* 2009; 78: 81–90

8.   Fox News Poll, 2011, Question 29. Available at: http://www. foxnews.com/us/2011/09/07/fox-news-poll-creationism/

9.   This quote and the one in the previous paragraph are from a 2005 interview with Sheri Kaplan published by TheBody. com, available at: http://www.thebody.com/hivawards/ winners/skaplan.html

      The biographical information given in this section comes from that article as well as two others: Cheakalos, C. 'Positive Approach: Sheri Kaplan gives heterosexuals with HIV a place to celebrate the joys of life', *People* magazine, 4 March 2002. Available at: http://www.people.com/people/archive/ article/0,,20136502,00.html

      Bradley Hagerty, B. 'Can Positive Thoughts Help Heal Another Person?', *NPR*, 21 May 2009. Available at: http:// www.npr.org/templates/story/story.php?storyId=104351710

      I was unable to contact Sheri to find out how she is doing now.

10.   *Spiritual Transformation and Healing: Anthropological, Theological,*

Neuroscientific and Clinical Perspectives. Koss-Chioino, J. & Hefner, P. J. (eds), AltaMira Press (2006), p. 245 (Sheri is named in this paper as 'Susan').

11. Cotton, S. et al. *Journal of General Internal Medicine* 2006; 21: S5–13

12. Ironson, G. et al. *Journal of General Internal Medicine* 2006; 21: S62–68

13. Sloan, E. et al. 2007. 'Psychobiology of HIV infection.' In Ader, R. (ed.), *Psychoneuroimmunology.* Academic Press, San Diego, pp. 869–895
    Cole, S.W. *Psychosomatic Medicine* 2008; 70: 562–568

14. Leserman, J. et al. *Psychological Medicine* 2002; 32: 1059–1073

15. Carrico, A.W. & Antoni, M.H. *Psychosomatic Medicine* 2008; 70: 575–584
    Creswell, J.D. et al. *Brain, Behavior and Immunity* 2009; 23: 184–188

16. Telephone interview with Andrew Newberg, 10 March 2014.

17. Pargament, K.I. et al. *Archives of Internal Medicine* 2001; 161: 1881–1885

18. Ironson, G. et al. *Journal of Behavioral Medicine* 2011; 34: 414–425

19. Ironson, G. et al. *Journal of Behavioral Medicine* 2011; 34: 414–425

20. Wachholtz, A.B. & Pargament, K.I. *Journal of Behavioral Medicine* 2005; 28: 369–384

21. Wachholtz, A.B. & Pargament, K.I. *Journal of Behavioral Medicine* 2008; 31: 351–366

22. Telephone interview with Kenneth Pargament, 12 March 2014.

23. Wachholtz, A.B. and Pargament, K.I. *Journal of Behavioral Medicine* 2005; 28: 369–384

24. Pargament, K.I. & Mahoney, A. *The International Journal for the Psychology of Religion* 2005; 15: 179–198

25. Jacobs, T.L. et al. *Psychoneuroendocrinology* 2011; 36: 664–681

26. Telephone interview with Clifford Saron, 4 April 2014.

27. This quote previously appeared in 'How Meditation Might Ward Off the Effects of Ageing' by Jo Marchant, *Observer*, 24 April 2011. Available at: http://www.theguardian.com/life andstyle/2011/apr/24/meditation-ageing-shamatha-project

28. Fredrickson, B.L. et al. *Proceedings of the National Academy of Sciences* 2013; 110: 13684–13689
    Marchant, J. 'The Pursuit of Happiness', *Nature* 2013; 503: 458–460

29. Cacioppo, J. & Patrick, W. *Loneliness: Human Nature and the Need for Social Connection* (2008), p. 262

30. Interview with Alessandro de Franciscis, Lourdes Medical Bureau, 12 June 2015.

31. This quote is taken from a talk given by Vittorio Micheli at Our Lady of Lourdes Church, Dublin, 23 May 2014.

32. Interview with Tim Briggs, Royal National Orthopaedic Hospital, Stanmore, Middlesex, 16 January and 20 February 2015.

Conclusion

1. 'Lending a hand that heals', King5, 16 September 2014. Available at: http://www.king5.com/story/entertainment/ television/programs/evening-magazine/2014/09/16/lending-a-hand-that-heals/15740091/
   For more information about Mary Lee McRoberts and her work, please see: http://www.maryleemcroberts.com/

2. While poorly designed studies sometimes show that patients benefit from reiki, once you do high-quality trials, in which reiki is compared against fake therapy, the benefits disappear. Edzard Ernst and his colleagues carried out a systematic review of RCTs in 2008 (Lee, M.S. et al. *The International Journal of Clinical Practice* 2008; 62: 947–954). In general, these

trials showed that real reiki worked no better than sham reiki. There were a few positive results for reiki, but these tended to be one-offs, where a particular benefit might appear in one trial but was not replicated in other trials. Most of these studies had flaws, such as being too small, being poorly designed, or that the data were not adequately reported. The authors concluded that 'the value of reiki remains unproven'.

3. One of the most rigorous analyses of this therapy was published in 2005 (Shang, A. et al. *The Lancet* 2005; 366: 726–732). It included 110 homeopathy RCTs and compared these to 110 equivalent trials of conventional medicines. When the authors restricted their analysis to the 'high-quality' trials, the conventional medicines were clearly better than placebo, whereas the homeopathic remedies showed only marginal benefit, consistent with them being no different to placebo (especially when you take into account that positive trials are more likely to be published than negative ones).

There have been other meta-analyses and systematic reviews of homeopathy trials, but none has ever shown convincing evidence that it works better than placebo. Nor have scientists ever been able to find any measurable difference between homeopathic remedies and inert liquids or pills.

4. Abbot, N.C. et al. *Pain* 2001; 91: 79–89
   Ernst has now retired, and is an emeritus professor of complementary medicine at Exeter University. For more information about his work, see http://edzardernst.com

5. Ernst, E. 'Running on faith', *The Guardian*, 15 February 2005. Available at: http://www.theguardian.com/society/2005/feb/15/health.medicineandhealth1

6. See, for example, the German New Medicine website on breast cancer: http://www.newmedicine.ca/breast.php

7. Several families claim that their relatives have died after refusing conventional treatment on Ryke Hamer's advice, for example, see: http://www.ariplex.com/ama/amamiche.htm

Deaths resulting from alternative care advised by other doctors include:

Sheldon T. 'Dutch Doctor Struck Off for Alternative Care of Actor Dying of Cancer', *British Medical Journal* 2007; 335: 13

'Alternative Cure Doctor Suspended', BBC News, 29 June 2007. Available at: http://news.bbc.co.uk/1/hi/england/london/6255356.stm

8. Schmidt, K. & Ernst, E. *British Medical Journal* 2002; 325:597

9. Jones, M. 'Malaria Advice "risks lives"', *Newsnight*, BBC2, 13 July 2006

10. For example, see:
    Kent, G.P. *American Journal of Epidemiology* 1988; 127: 591–598
    Ernst, G. et al. *Complementary Therapies in Medicine* 2003; 11: 93–97

11. McRoberts responds that she's confident the spirits she communicates with would not show her anything that might be harmful to a patient. 'My information comes directly from the other side,' she says, 'and I totally trust that it's exactly the way it is supposed to be. If I were using my brain to think of what to do with the client, it would be another matter. But I turn my brain off when I connect in and let them feed me directly.' Email from Mary Lee McRoberts, 29 August 2015.

12. For a discussion of the history and mechanism of acupuncture, see: Singh, S. & Ernst, E. *Trick or Treatment* (2008), Chapter 2, pp. 39–88.

13. For most complaints, there is no evidence in high-quality trials that acupuncture works better than placebo. However for certain types of chronic pain and nausea, it may have a physical effect as well as a psychological one. A 2012 systematic review of 29 trials for chronic pain including 17,922 patients (Vickers, A.J. et al. *Archives of Internal Medicine*

2012; 172: 1444–1453) found that real acupuncture works slightly better than sham acupuncture (and both work better than a no-acupuncture control). The authors concluded that although most of the benefit of acupuncture is a placebo effect, the needles may have a modest effect too.

14. Interview with Deming Huang, Stanford Center for Integrative Medicine (SCIM), Stanford, California, 26 November 2013.

15. Freedman, D.H. 'The Triumph of New-age Medicine', *The Atlantic*, July/August 2011. Available at: http://www.theatlantic.com/magazine/archive/2011/07/the-triumph-of-new-age-medicine/308554/

16. Interview with Jeremy Howick, Oxford, 20 April 2015.

17. Stroud, L.R. et al. *Biological Psychiatry* 2002; 52: 318–327
    Kudielka, B.M. et al. *Biological Psychology* 2005; 69: 113–132

18. Email interview with Elissa Epel, 9 April 2015.

19. Telephone interview with Jeff Sloan, 25 February 2015.

20. See also Sloan's work with quality-of-life measures:
    Frost, M.H. & Sloan, J.A. *The American Journal of Managed Care* 2002; 8: 5574–9
    Sloan, J.A. et al. *Journal of Clinical Oncology* 2012; 30: 1498–1504

21. Heathcote, E. *British Medical Journal* 2006; 333: 1304–1305

22. UCSF's Thomas Bodenheimer estimated it at 70% in 2000 (Bodenheimer, T. *New England Journal of Medicine* 2000; 342: 1539–44). Harvard's John Abramson, author of the 2004 book *Overdosed America*, says that by 2009 this figure had reached 85%. See: http://www.ourbodiesourselves.org/health-info/who-paid-for-that-study/

23. The annual budget of the National Center for Complementary and Integrative Health in 2015 was $124.1 million (0.4% of the NIH annual budget of $30 billion). I wasn't able to find an exact figure for how much of this is spent on trials of mind–body therapies, but according to the centre's third strategic plan (2011–2015), the money is split between two

main research areas – mind–body therapies and natural products. Some of the money also goes on things like studying how many people use complementary and alternative medicine (CAM), and disseminating evidence-based information on CAM interventions.

See: https://nccih.nih.gov/sites/nccam.nih.gov/files/about/plans/2011/NCCAM_SP_508.pdf

24. Shang, A. et al. *The Lancet* 2005; 366: 726–732

The authors included 110 homeopathy RCTs and compared these to 110 equivalent trials of conventional medicines. Twenty-one of the homeopathy trials were judged to be of 'high quality', compared to just nine of the conventional trials.

25. Skype video interview with Elvira Lang, 24 April 2014.

26. Telephone interview with Ellen Hodnett, 10 March 2014.

27. Interview with Bill Eley, Emory University in Atlanta, Georgia, 5 February 2015.

28. At least 400 US physicians commit suicide every year (equivalent to losing a whole medical school); double the risk faced by the general population.

Andrew, L.B. et al. 'Physician Suicide', *Medscape* 2014. Available at: http://emedicine.medscape.com/article/806779-overview

Young doctors are especially vulnerable, with problems starting in school. In a 2009 study, nearly 10% of fourth-year medical students and interns admitted to having suicidal thoughts in past two weeks.

Goebert, D. et al. *Academic Medicine* 2009; 84: 236–241

Burnout – a psychological syndrome that includes emotional exhaustion and depersonalisation – is estimated to affect as many as half of medical students, and more than a third of physicians.

Hojat, M. et al. *International Journal of Medical Education* 2015; 6: 12–16

Recent research suggests that loss of empathy for patients may be a contributing factor in burnout. In brain imaging studies, doctors in general have less empathy-related brain activity than others when viewing photos of people in pain, and the lowest levels of empathy-related brain activity are associated with more severe burnout.

Tei, S. et al. *Translational Psychiatry* 2014; 4: e393

29. In 2013, the US spent $2.9 trillion on healthcare, or 17.4% of GDP, see: http://www.cms.gov/Research-Statistics-Data-and-Systems/Statistics-Trends-and-Reports/NationalHealth-ExpendData/downloads/highlights.pdf

    For comparison with other countries, see: http://data.world-bank.org/indicator/SH.XPD.TOTL.ZS

30. See: http://www.cdc.gov/nchs/fastats/drug-use-therapeutic.htm

    Also, Thompson, D. 'Prescription Drug Use Continues to Climb in US', WebMD News, 14 May 2014. Available at: http://www.webmd.com/news/20140514/prescription-drug-use-continues-to-climb-in-us

31. Budnitz, D.S. et al. *New England Journal of Medicine* 2011; 365: 2002–2012

32. Schork, N.J. *Nature* 2015; 520: 609–611

33. Gøtzsche, P.C. *British Medical Journal* 2015; 350: h2435

34. James, J.T. *Journal of Patient Safety* 2013; 9: 122–128

    For statistics on leading causes of death, see: http://www.cdc.gov/nchs/fastats/leading-causes-of-death.htm

35. http://www.fda.gov/Drugs/DevelopmentApprovalProcess/DevelopmentResources/DrugInteractionsLabeling/ucm-114848.htm

    These figures date from 2000, so it may be significantly more than that by now.

36. See Young, E. *SANE: How I Shaped Up My Mind, Improved My Mental Strength and Found Calm* (2015) for a fascinating and evidence-based exploration of how physical factors, such as diet, exercise and sleep, influence the mind.

ACKNOWLEDGEMENTS

While reporting this book, I've been impressed and touched by the generosity of those who have spent time answering my questions and sharing with me their ideas and experiences. The final result, *Cure*, would not exist without the expertise, patience and support of countless individuals, and I hope that it lives up to the trust they have placed in me.

First, thank you to the scientists and medical professionals who took time out of their busy schedules to explain their work to me, and invited me into their labs and consulting rooms. I am particularly grateful to Fabrizio Benedetti for welcoming me to Plateau Rosa; Elisa Frisaldi and Elisa Carlino for allowing me to watch their experiments at the Molinette Hospital in Turin; Ted Kaptchuk and Nicholas Humphrey for sharing their views on placebos in Cambridge, Massachusetts, and Cambridge, UK, respectively; Manfred Schedlowski and his team at the University of Essen for letting me taste the famous green drink; and Peter Whorwell and Pamela Cruickshanks for introducing me to their patients in Manchester.

I'm similarly indebted to David Patterson, Sam Sharar, Christine Hoffer, Hunter Hoffman and all at Harborview for showing me the potential of virtual worlds; Elvira Lang at Hypnalgesics plus

Kelly Bergeron and Pamela Kuzia at Boston Medical Center for letting me witness Comfort Talk in action; Patricia Saintey at Heartfelt Consulting for demonstrating heart rate variability biofeedback; Steve Cole for putting up with multiple interviews and showing me around at UCLA; Lobsang Negi, Bill Eley, Brendan Ozawa-de Silva, Samuel Fernandez-Carriba, Jennifer Mascaro and especially Timothy Harrison for introducing me to CBCT. Huge thanks also to Michael Moran and colleagues for allowing me the privilege of visiting and volunteering at Lourdes. I was blown away by your compassion and commitment.

Many others gave their time and expertise while I was reporting this book, and articles that fed into this book. They include Jerry Jarvik, David Kallmes, David Spiegel, Sara Lazar, Alessandro de Franciscis, Jon Stoessl, Dan Moerman, Irving Kirsch, Edzard Ernst, Adrian Sandler, Karen Olness, Oliver Witzke, Tim Noakes, Chris Beedie, Peter White, Elizabeth Blackburn, Elissa Epel, Jue Lin, Edoardo Casiglia, Enrico Facco, Candy McCabe, Ellen Hodnett, Vicki Jackson, Jennifer Temel, Robert Kloner, Mary Armanios, Gene Brody, Greg Miller, Wendy Mendes, Paul Lehrer, Barbara Fredrickson, Bethany Kok, Richard Sloan, Andrew Newberg, Kenneth Pargament, Clifford Saron, Olive Conyers, Tim Briggs, Mark Williams, Giuseppe Pagnoni, Trudy Goodman, Christiane Wolf, Willem Kuyken, David Gorski, Robert Simpson, David Rehkopf, Michel Poulain, John Cacioppo, Michelle Carlson, Charles Raison, James Coyne, Michael Antoni, Simon Norburn, Bonnie McGregor, Mary Lee McRoberts, Catherine Mayer, Jeremy Howick, Ben Goldacre, Jeff Sloan, Tom Stannard, Kavita Vedhara, Gaëlle Desbordes, Jacqui Tomkins, Dan Martin, Michael Irwin, Helen Lavretsky, Clare Stevinson and Marc Schoen.

I started this book fascinated by the science of how our minds might influence our bodies, but speaking to patients and trial volunteers helped me to realise that, beyond its intellectual importance, this subject has profound practical consequences for our health and how we all live our lives. For me, their stories bring this book alive. They include Bonnie Anderson, Rosanna Consonni, Linda Buonanno, Simon Bolingbroke, Karl-Heinz Wilbers, Samantha Miller, Gareth Walker, Lupita Quereda, Rose Wise, Caroline Dempsey, John Flynn and Tunde Balogh. There are others whom I have not named to protect their privacy, and many more whose words are not included in these pages – they all informed this book and I'm greatly indebted to each of them.

This book began as a feature article in *New Scientist* magazine. Thanks to Michael Le Page there for not only saying yes to my idea but putting it on the cover, and to all the editors who have worked with me on related articles since, including Mun-Keat Looi at Mosaic and all at *Nature*. I'm grateful to Kevin Fong, Mark Henderson and Niki Jakeways for finding time to read the finished draft, and offering thoughtful and helpful comments. And thanks to Gaia Vince and Emma Young for keeping my stress levels down with friendship, advice and adventures, including the discovery of the best spa in the world.

My brilliant agent, Karolina Sutton, believed in this book from the beginning and provided invaluable comments on my proposal and throughout the writing process. Thanks also to my lovely copy editor, Octavia Reeve, and to my editors Amanda Cook at Crown and Katy Follain at Canongate, for seeing what this book could be and pushing me to achieve that potential. I'm so grateful to have the opportunity to work with such patient and talented people.

Finally, thank you to my family: to Ian Sample, my partner and best friend, for rock-solid encouragement and support; and to Poppy and Rufus, my beautiful children, for joy, hugs and all the inspiration I'll ever need.

INDEX

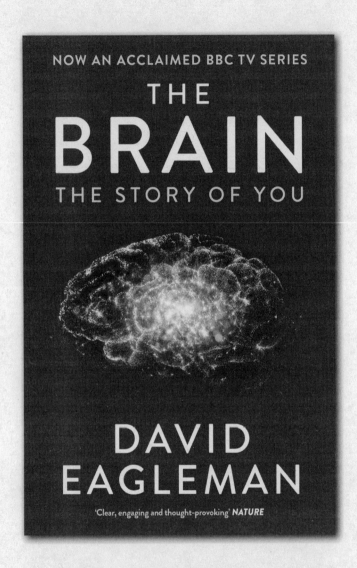

NOW AN ACCLAIMED BBC TV SERIES

THE
BRAIN
THE STORY OF YOU

DAVID
EAGLEMAN

'Clear, engaging and thought-provoking' *NATURE*

'Entertaining and profound: page-turning
neuroscience from a bit of a genius' *Guardian*

CANON‖GATE

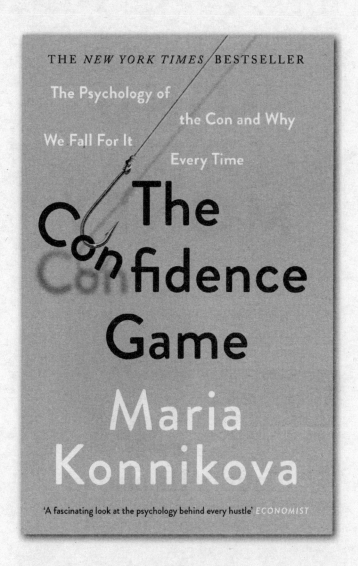

The Psychology of
the Con and Why
We Fall For It
Every Time

The Confidence Game

Maria Konnikova

'A fascinating look at the psychology behind every hustle' *ECONOMIST*

'Remarkable . . . *The Confidence Game* will widen your
eyes and sharpen your mind' Daniel H. Pink

CANON GATE

'Hugely enjoyable' OLIVER BURKEMAN

MIND
OVER
MONEY

The
Psychology
of Money
and
How to Use
It Better

Claudia Hammond

Presenter of BBC Radio 4's *All in the Mind*

'Interesting and insightful'
Sunday Times

CANON❙❙GATE